World Geography Mysteries

Written by Mark Falstein ◆ Illustrated by Kelly Kennedy

The Learning Works

The purchase of this book entitles teachers to reproduce copies of the activities in this book for use in their classroom. The reproduction of any part for an entire school or school system or for commercial use is strictly prohibited.

LW 382
ISBN: 0-88160-261-2

© 1998—The Learning Works, Inc.
Santa Barbara, CA 93160
All rights reserved.
Printed in the United States of America.

Contents

To the Teacher 5–6

Exploring the Western Hemisphere

Spying Out the U.S.A. 8–10

Canada, East to West 11–13

Deeper Into Mexico 14–16

Mystery Cruise 17–19

Down South America Way 20–22

Exploring the Eastern Hemisphere

The Countess's Grand Tour 24–26

Desert Lands 27–29

African Safari 30–32

Half the World is Asia 33–35

Half of Asia is Southern Asia 36–38

Lands Down Under 39–41

Deeper Into the World

Mountains and Deserts 43–45

Lakes and Inland Seas 46–47

Around the World, Island to Island 48–50

New on the Map 51–53

Follow the River 54–56

Dead Reckoning 57–58

Contents
(continued)

The Complete World Traveler

A Capital Adventure 60–62

Don't Forget Your Camera! 63–65

For Those Who Love Adventure 66–68

Natural Wonders 69–71

Celebrity Landmarks 72–74

Wildlife Scavenger Hunt 75–77

A World Full of People

The World of Music and Dance 79–81

The World of Fun and Games 82–84

Sid and Max, Together Again 85–87

Let's Party! 88–90

Answer Key 91–96

 # To the Teacher

World Geography Mysteries is designed to acquaint students in grades 4 through 6 with world geography in an engaging and entertaining manner. Each of these two- or three-page activities contains geographical clues to 10 locations, which students must find with the help of a map, globe, or atlas. The locations may be large or small cities, countries, or geographical features. Where a location appears in more than one activity in the book, different geographical clues are used in each instance to help the students locate it. Both customary and metric units are used in the clues.

In the easiest activities, clues are independent of one another, and each one points to a specific location. Other activities contain progressive clues where the student must correctly identify one place name in order to use the next clue.

Map skills, in particular the skills of interpreting directions, decoding symbols, and using a scale, are the chief objectives of *World Geography Mysteries*. As the puzzles often require students to resolve ambiguities, apply deduction, and synthesize information, the activities will stimulate their higher-level thinking skills as well.

The primary clues are typically given in terms of direction and distance from an easily found map location, *e.g.*, "About 200 miles (322 km) southeast of Paris . . ."; "About 615 miles (990 km) south and a little west of Beijing . . ." "About 360 miles (580 km) due north of Mexico City." Most clues are deliberately unspecific in this manner, as precise distances are all but impossible to determine with a standard map scale, and precise directions can be conveyed only by the use of degree headings. Accordingly, secondary clues are given: locations relative to bodies of water, state borders, and other nearby map features. In some cases, where students are asked to find a location among many place names within a relatively small area, a tertiary clue is given, *e.g.*, "The name of the city has seven letters, beginning with 'A.'" Most clues contain an interesting or unusual fact about the location as well. Place and feature names used in *World Geography Mysteries* can be found in most standard atlases. All can be found in the maps included in country articles in *World Book Encyclopedia*.

The prospect of solving a mystery or puzzle will provide students with added motivation. In many activities, the mystery involves finding a lost fortune or document, or pursuing a criminal. In others, the only "mystery" is the information the student is required to find through thought and research.

 # To the Teacher
(continued)

Contents

The book is divided into five sections, though the activities need not be presented in any particular order:

Exploring the Western Hemisphere: Each of these five activities focuses on a specific country, subcontinent, or continent: the United States, Canada, Mexico, Central America and the Caribbean, and South America.

Exploring the Eastern Hemisphere: These six activities continue the region-by-region survey of the globe, focusing respectively on Europe; North Africa and Southwest Asia; Sub-Saharan Africa; Northern Asia; Southern Asia; and Australia, Oceania, and Antarctica.

Deeper Into the World: In these six activities, the clues ask students to identify mountain ranges and deserts, inland bodies of water, islands, and recently created nations. One activity in this section calls on students to identify locations by their latitude and longitude coordinates.

The Complete World Traveler: These six activities focus on world capitals, popular and more adventuresome tourist destinations, natural wonders, famous landmarks, and wildlife preserves. In the first activity in this section, "capital" refers not necessarily to a seat of government but to a city that has primacy in some historical or cultural way.

A World Full of People: These last four activities call on students to identify geographic locations associated with music and dance, sports, works of juvenile literature, and festivals.

How to Use This Book

As a whole-class activity: Students can solve a World Geography Mystery as a group, with individuals being called upon to solve the clues using a wall map or pull-down map. If detailed country and world-regional maps are available in student geography texts or atlases, students may be assigned an activity as desk work. Completed activity sheets may be placed in students' file folders or portfolios.

As a learning-center activity: Small groups of students may be given an activity to complete cooperatively at a learning center, using an atlas or other appropriate reference. As an added motivation, you might time various groups on successful completion of an activity, and devise a point system for order of finish for a class competition.

As an extra-credit or homework activity: Students can be assigned activities to complete using available references at home, or in the school or public library. Completed activity sheets may be placed in students' file folders or portfolios.

Spying Out the U.S.A.

Something was wrong with my computer. I had located the source of the signal that showed me where the kidnappers were holding the president's little dog. But all I had was a satellite map, not a place name. The satellite image showed a great, crooked Y with a large round structure of some sort where the three lines met. I set off in my trusty rocket plane to find the place.

1. Leaving my home town of Detroit, Michigan, I flew about 235 miles (378 km) southwest. I was over a big city where the White River cuts across the broad Midwestern plain. In the west of the city was a motor speedway. They hold a famous car race there every May. But it was more rectangular than round, and I saw no Y.

2. I headed south, to a city whose name suggests its location—straddling a state line. It's about 335 miles (540 km) northwest of New Orleans, Louisiana. North of the city, the Red River forms part of the boundary between the two states. But I could see at a glance that this was not the place.

3. I next flew to the Pacific Northwest, to a port city about 30 miles (48 km) south of Seattle and about 125 miles (201 km) north and a little east of Portland. I could see fishing boats in Puget Sound, and the volcanic cone of Mount Rainier to the southeast. But that was the only "hot spot" around.

4. I flew diagonally across the country, to a city on a bay of the Gulf of Mexico. It has the same name as a great city in Russia. It's about 450 miles (720 km) southeast of Birmingham, Alabama, and only about 335 miles (540 km) north of Havana, Cuba. But I saw no sign of any dognappers, foreign or domestic.

5. I traveled next to a city on another bay. The bay has the same name as its state. The city is about 190 miles (306 km) northeast of New York City. There were roads radiating out in all directions to all the New England states. That must be why they call this city "the hub." The pattern was not so much a Y as a collection of Ys. You know what they say about a word to the Ys, don't you?

6. I crossed the country again, to a town on the Russian River that's about 105 miles (169 km) northwest of San Francisco, and about 180 miles (290 km) west of the capital city of Nevada. Its name has five letters, beginning with U, but no Y. I was beginning to ask why myself.

7. I was really burning up rocket fuel! My next search point was on the Mississippi River, just across from Davenport, Iowa, and about 170 miles (274 km) southwest of Milwaukee, Wisconsin. It's where the Rock River joins the Mississippi, and I suppose it was an island in that river which gave the town its name. I saw a lot of railroad tracks, some of which formed Ys, but nothing on the scale shown by the satellite photo.

Spying Out the U.S.A.

8. There were still some regions of the country where I hadn't looked. One was the desert Southwest. Evening found me circling over a town whose name sounds like something a horse does. A lot of Navaho people live in the town. It's less than 25 miles (40 km) southeast of the Navaho capital of Window Rock, Arizona, and about 300 miles (483 km) southwest of Pike's Peak, in Colorado. There were rocks there, but I still couldn't find my "quarry."

9. Desperate now, I flew back east. I crossed the Ohio River at Cincinnati and proceeded to a city on that river that's about 125 miles (201 km) southeast of Cincinnati, and about 130 miles (208 km) east and a little north of the capital of Kentucky. The city is not in Kentucky, or in Ohio, either.

10. There had to be a pattern here—and there was. When I listed the places I had been and noted their initial letters, it came to me. All I needed was the first letter to complete the pattern. From the last place, I flew a little less than 200 miles (322 km) northeast to a city where two rivers meet to form the Ohio—and there was the Y, formed by the junction of the rivers! And there was the round structure: a stadium! There was no doubt about it—I had to get a new computer.

Can you identify the 10 places suggested by the clues? Write the city and town names and state abbreviations on the lines below.

1. _____ 6. _____
2. _____ 7. _____
3. _____ 8. _____
4. _____ 9. _____
5. _____ 10. _____

World Geography Mysteries
© The Learning Works, Inc.

Canada, East to West

They call me Malamute Maggie. I'm returning from an around-the-world flight in my private plane. I live in a cold place, but I'm just *steamed* at how little folks out there know about my country. No, I'm *not* from the United States! I'm from that very large country just north of there. That's right—Canada: land of mystery. Come along with me as I complete my journey by traveling across my country. We'll have fun learning a thing or two as we go.

1. We start in a city on the Atlantic, a little more than 300 miles (483 km) east of Augusta, Maine. It's in a province (that's like a state) whose name means "New Scotland." It is Canada's busiest Atlantic port and has the country's biggest naval base. Its name ends with the letter X.

2. Now we wing almost 400 miles (644 km) northwest, across the Bay of Fundy, the province of New Brunswick, and the U.S. state of Maine. We're going to Canada's oldest city, at a place where the St. Lawrence River narrows. In fact, the name means "place-where-the-river-narrows" in the Huron language, though a lot of people think it's French. *Parlez-vous français?* It'll come in handy here.

3. Let's continue southwest up the St. Lawrence, about 145 miles (233 km) to Canada's largest city. They speak French here too—it's the world's largest French-speaking city except for Paris. If we had time to stop, we could check out the "underground city"—a huge mall that lies underneath the city's downtown area.

4. Now we fly about 100 miles (161 km) west to Canada's capital. It's on the border between Ontario and Quebec, the two largest provinces. Its name comes from an Algonquian Indian word that means "to trade." You can probably guess what activity was prominent here about 175 years ago.

5. Our next hop takes us about 230 miles (370 km) southwest to a city on Lake Ontario. It's our second-largest city, but our largest metropolitan area and business center. Whoops, got to bank left here, or we'll hit the CN Tower. It's 553 meters tall—1,815 feet to you. Wave to the people in the restaurant as we fly by!

6. Now we head about 930 miles (1,496 km) northwest to the city we call our "Gateway to the West." It's in the province of Manitoba, about halfway between the Atlantic and Pacific oceans and about 60 miles (97 km) north of the U.S. border. Just to the north is a large lake with the same name as the city.

7. We're over Canada's Great Plains now—wheat country. We're flying about 330 miles (531 km) west and a little north to the largest city in the province of Saskatchewan, between Indian Head and Moose Jaw. The name of the city is the Latin word for "queen," a reminder that Canada was once a dominion of Great Britain.

8. From there we fly about 415 miles (668 km) west and a little north to our third-largest city, located on the Bow River. Western Canada is a lot like the western United States. This city is a center for the oil business and cattle ranching. It's also the home of the "Stampede," the most famous rodeo in the world.

9. Our next stop is a city about 415 miles (668 km) further southwest. We've reached the sea again—the Strait of Georgia, an arm of the Pacific. This city is Canada's busiest port and home to many people of Chinese descent. On the other side of the strait is a large island with the same name as the city.

10. I'm almost home now—well, it's not much more than another 900 miles (1,448 km) northwest. We're going to the Yukon Territory, to a city northwest of Carcross and south of Lake Laberge. Mining and trapping are still the most important businesses here, just as they were in Gold Rush times. It's still pretty wild and rugged—just like me! Thanks for joining me on my trip across my country.

Can you identify the 10 Canadian cities Maggie visited on her flight? Write their names on the lines below.

1. _____   6. _____
2. _____   7. _____
3. _____   8. _____
4. _____   9. _____
5. _____   10. _____

Deeper Into Mexico

Hotfoot Hotchkiss, international jewel smuggler, had escaped again. This time, he had crossed the border into Mexico. I had to track him down and bring him back to face justice. My mom always told me I should have paid more attention in Spanish class.

1. I started out where many *norteamericanos* do when they visit Mexico. It is a city in the northwestern corner of the country, just over the border from San Diego, California. Hotfoot had been spotted taking in the bullfights at the Bullring by the Sea. By the time I'd gotten through traffic, I had just missed him.

2. From there I headed about 715 miles (1,150 km) southeast across the Sonora Desert and the Western Sierra Madre to the capital of a Mexican state. It has the same name as the state. Hotfoot loves small dogs, and this city is famous for a particular breed. I visited a few kennels and soon picked up Hotfoot's trail.

3. I caught a plane southeast and flew about 575 miles (925 km) to the state of Jalisco. I landed in Mexico's second-largest city. It has a famous university where many people from the United States come to study. If Hotfoot had stayed in school, maybe he wouldn't have ended up on the wrong side of the law. The local police told me he'd been spotted buying scuba-diving equipment.

4. I traveled about 350 miles (563 km) southeast to a beach resort town. Its name has eight letters, beginning with "A." Sure enough, Hotfoot had been there. A lot of vacationing *norteamericanos* had seen him. So had the police. I stopped to watch the famous cliff divers making their spectacular leaps into the Pacific Ocean.

Deeper Into Mexico

5. Hotfoot knew I was on his trail. I figured he would try to lose himself in a crowd. I traveled about 185 miles (298 km) north and a little east to one of the world's largest cities. Once the site had been covered by a lake. On an island in that lake, the ancient Aztecs built their capital, Tenochtitlán. The city has many magnificent sights, but Hotfoot wasn't one of them.

6. It was the fifth of May—*Cinco de Mayo* to the locals. I headed about 65 miles (105 km) southeast, past the Iztaccihuatl volcano, to the city where the holiday began. It was there in 1867 that the Mexicans routed an invading French army. The festivities weren't much. It turns out that Cinco de Mayo is a bigger holiday among Mexican Americans in the United States than in Mexico. But I did spot Hotfoot heading eastward in a rental car.

7. Hotfoot might have been trying to leave Mexico by ship. I followed him about 135 miles (217 km) east and a little north to a port on the Gulf of Mexico. It was here that Cortés and his *conquistadores* landed in 1519 to begin their conquest of Mexico. Its name means "true cross" in Spanish. Hotfoot had truly crossed me. I found that he had left the city by plane, not by ship.

8. Amazingly, Hotfoot had filed a flight plan. I chartered a plane and followed him, about 620 miles (1,000 km) northeast across the Bahía de Campeche and the Yucatan Peninsula. In a resort town with a six-letter name beginning with "C," I learned that he had gone into the rain forest. "He is probably visiting the Mayan ruins, señor," a helpful hotel clerk told me.

•15•

Deeper Into Mexico

9. Three days later, I stumbled out of the jungle. I hadn't seen Hotfoot, but my own feet certainly needed cooling down. I traveled about 700 miles (1,127 km) southwest to an old Aztec city whose name is pronounced "wahocka." It's the capital of a state of the same name. I spent a few days resting and wondering what the chief would say when I told him I hadn't found Hotfoot.

10. Wouldn't you know—I caught him just as I was coming back into the United States! I had gone about 1,180 miles (1,900 km) northwest to a border city opposite El Paso, Texas. I spotted him going through the gate and grabbed him on American soil—North American soil, that is. "You're going to jail," I told him, "and by the way, chump, thanks for giving me the chance to see Mexico."

Can you identify the Mexican cities where the detective pursued Hotfoot? Write their names on the lines below.

1. _____   6. _____
2. _____   7. _____
3. _____   8. _____
4. _____   9. _____
5. _____   10. _____

Mystery Cruise

Good day, and welcome to our famous Quick Caribbean Cruise. We're going to visit 10 cities in 10 countries in approximately 10 minutes. Now, some of these places are not exactly in or on the Caribbean Sea. Some of them you can't even get to by boat. But a map of the Caribbean region will help you find them. See how many you can identify from the clues without getting lost at sea.

1. Let's start in a city where English is the native language. It's on New Providence Island, one of about 3,000 islands and coral reefs that make up this country. The city is on the 25th parallel of north latitude and about 185 miles (298 km) southeast of Miami, Florida. Enjoy the seafood and the scuba diving!

2. Now let's visit another island capital. It's about 350 miles (563 km) southwest of the first place we stopped and about 230 miles (370 km) southwest of Miami. Spanish colonists built the city in 1519, and many of its houses are more than 400 years old. Cigars made here are famous the world over, but they'll make you just as sick as cigars made anywhere else, so leave them alone!

3. Now let's travel almost 800 miles (1,287 km) southwest to the mainland of Central America. We're in a capital with the same name as its country, about 650 miles (1,046 km) southeast of *another* capital with the same name as its country. It's in the mountains, closer to the Pacific than to the Caribbean. Spanish is spoken here, but in the nearby countryside, many people speak Native American languages as well.

4. We're still on the Pacific Ocean side of Central America. We're in a highland city about 110 miles (177 km) southeast of the last place we visited and about 135 miles (217 km) southwest of Tegucigalpa, Honduras. Many people from around here emigrated to the United States during a civil war in the 1980s.

5. We continue on southeast about 225 miles (362 km). We cross the Golfo de Fonseca and arrive at a city on the south shore of a lake. The lake and the city have the same name. The name of the country comes from the Nicarao, a Native American people whose cities stood here when the Spanish arrived in 1522.

6. We make our last Central American stop about 510 miles (821 km) farther southeast. We're in a bustling seaport. The city, the country, and the body of water we're facing all have the same name. We're looking out over the Pacific, but the Caribbean is only a few hours from here by ship.

7. Let's return to the islands. We travel about 645 miles (1,038 km) north and little east of our last mainland stop. Our landing is a seaport and capital city about 830 (1,335 km) northwest of Caracas, Venezuela. Is your English better than your Spanish? Then you're in luck, because English is the official language here. Welcome to the home of reggae music!

Mystery Cruise

8. Now we sail about 295 miles (475 km) east and a little north. We're in another seaport capital, and French is spoken here. This country was founded about 200 years ago by former slaves who overthrew their masters. Mexico City is about 1,715 miles (2,759 km) almost due west.

9. Next, we travel about 160 miles (257 km) east and a little south. We're on the same island, but in the capital of a different country. Are you a baseball fan? More major-league players come from this country than from any other country except the United States.

10. Finally, we head back to the U.S.—almost. This island is a United States dependency, and one day it may become a state. We're in the capital city, about 250 miles (402 km) east of the last place we stopped—and about 1,640 miles (2,639 km) southeast of New York City, where many of its former citizens now live.

Can you identify all of the cities and countries you visited on the cruise? Write their names on the lines below.

1. _____   6. _____
2. _____   7. _____
3. _____   8. _____
4. _____   9. _____
5. _____   10. _____

Down South America Way

I found the old piece of goatskin in a library in Lima, Peru. It looked like a map with a few lines scratched on it. For years I'd been searching for a clue to Atahualpa's treasure. This vast store of gold was rumored to have been hidden by the last Incan emperor at the time of the Spanish conquest of Peru. Of course, it was only a legend. But the Incan trade routes had reached all over South America—and so, it appeared, did this map. I would have to follow what looked like a lot of marked routes.

1. I started my search in the old Incan capital. It stands about 11,200 feet (3,400 m) high in the Andes Mountains, about 354 miles (570 km) southeast of Lima and about 185 miles (298 km) northwest of Lake Titicaca. The Incan language, Quechua (KETCH-wa) is still spoken there, along with Spanish.

2. Following the map, I struck off southeastward to a city a little more than 35 miles (56 km) east of Lake Titicaca. I was in a different country now. I was in one of its two capital cities and about 260 miles (418 km) northwest of the other. At about 12,000 feet (3,660 m) above sea level, it's the highest capital in the world. I had nowhere to go but down.

3. I cut back across the Andes to the Pacific coast. I was headed for a busy port where I could charter a boat. The port is located just two degrees south of the equator, in a country whose name means "equator." It's about 170 miles (274 km) southwest of that country's capital, another old Incan city. I found a boat and set out, map in hand, for my next port of call.

4. I sailed north through the Panama Canal and eastward along the coast of Colombia. Rounding Punta de Gallinas, I put into a large bay. I docked near the head of the bay, on the west side. The harbor was full of oil tankers and drilling platforms. I was in this country's second-largest city, about 325 miles (523 km) west of the capital, Caracas.

5. After consulting with a local scholar, I was back at sea. My next stop wasn't in an independent country. It was in a dependency of France, the last European colony in South America. I was in its only large city, about 1,490 miles (2,400 km) east of Bogotá, Colombia, and about 425 miles (684 km) southeast of Georgetown, Guyana. The city has given its name both to a type of spicy hot pepper and to a pineapple.

6. I didn't know whether I was near the treasure, but I did know that I needed a rest. I sailed down the continent's east coast. When I reached Guanabara Bay, I was in South America's second-largest city. It lies about 220 miles (354 km) east and a little north of South America's largest city, São Paulo, Brazil. With Sugar Loaf Mountain rising above the bay, I knew I was in for a nice vacation.

7. A line on the map directed me to a city about 535 miles (861 km) north and a little west of São Paulo, and about 1,335 miles (2,150 km) east of La Paz, Bolivia. Now I was puzzled. This capital didn't exist in Incan times. In fact, it wasn't built until the 1950s! Could the map be—I hardly dared to ask myself—a fake?

8. I compared the map to others I found in a library in South America's third-largest city. This city, which lies on the Rio de la Plata, is located approximately 125 miles (201 km) west of Montevideo, Uruguay. After careful study, I realized I'd been wasting too much time at sea. I decided that the thing to do was to go back to the Andes.

9. I flew west and a little north, crossing the mountains. I landed in the capital of another country, about 65 miles (105 km) southeast of the port of Valparaíso. There I showed the map to a university professor. "Take it to my brother," she said, scribbling an address. "He will know what it means."

10. Her brother lived in the world's southernmost city. It's on the Strait of Magellan, at 53° south latitude. It's about 1,365 miles (2,197 km) south and a little east of its country's capital, and a little more than 800 miles (1,287 km) from the coast of Antarctica. It was worth the trip. The professor's brother turned out to be an automobile mechanic. "That's not a map; it's an oil rag," he said. "Those lines were made by a dipstick, sure as I'm standing here." I felt a bit like a dipstick myself.

Can you identify the cities and countries the treasure seeker visited? Write their names on the lines below.

1. _____ 6. _____
2. _____ 7. _____
3. _____ 8. _____
4. _____ 9. _____
5. _____ 10. _____

Exploring the Eastern Hemisphere

A Number of Heroes

The Countess's Grand Tour

In the glittering salons of Europe, she is known as Countess Eleanor-Rosine von Zwickau de la Tour, but the crime fighters of the international police agency STOP simply call her Madame—the most fearless agent of them all. We join Madame in the company of Jules Mazzini-Romanov, international jet-setter and suspected jewel thief. Will she put a stop to his life of crime?

1. Madame joins Jules *en route* from New York to London, England. Is there a reason he wants to stop in one of Europe's outposts on the way? This national capital started as a Viking settlement. It lies about 1,165 miles (1875 km) northwest of London, and about 980 miles (1,577 km) northwest of Trondheim, Norway. This is the "Land of the Midnight Sun." But so far, Madame is in the dark.

2. At Heathrow Airport, Jules immediately books a flight to a city about 290 miles (467 km) northwest of London, and about 125 miles (201 km) west of Liverpool, England. This city on the Shannon River was once ruled from London, but it has been the capital of a free nation since 1919. Madame trails Jules from the hotel, but finds no evidence that will end *his* freedom.

3. Back in England, Jules and Madame travel by train through the English Channel Tunnel. Their destination: the City of Light. To Madame and many others, it is the center of the world. It's about 210 miles (338 km) southeast of London, and about 250 miles (402 km) west of the German border. Madame and Jules attend a *soirée* at the Ritz Hotel and dance beside the Seine River. Madame is disappointed that Jules doesn't try to steal her ruby necklace.

The Countess's Grand Tour

4. Where is Jules off to next? To the sunny South, and a city by the Mediterranean Sea. It is about 185 miles (298 km) southeast of its country's capital, Madrid, and about 400 miles (644 km) southwest of Marseilles, France. Jules meets a mysterious man in an 800-year-old Moorish house. While Madame watches from a doorway, she eats one of the oranges for which the city is famous.

5. Jules and Madame are on the move again. They fly to a city about 70 miles (113 km) northeast of Lyon, France. French is spoken here, although the city is not in France. Their hotel is on a lake with the same name as the city, though the French call it "Lac Leman." The morning they leave, Madame reads a headline in a local newspaper: "Jewelry reported missing from six of the city's finest mansions."

6. Their destination is a large city south of the Alps. It lies almost 300 miles (483 km) northwest of Rome, Italy, and about 325 miles (523 km) west and a little south of Zagreb, Croatia. Madame and Jules take in the opera at La Scala. During the second act, Jules leaves. He returns without explanation just before the opera ends. "Get your coat—we're leaving," he whispers urgently.

7. They take the night train over the Alps. Morning finds them in a city a little more than 300 miles (483 km) southwest of Berlin, its country's capital, and about 220 miles (354 km) west and a little south of Vienna, Austria. Jules says he has urgent business that will keep him busy all day. Madame says she plans to visit the city's famous museum of technology. Instead, she follows him.

The Countess's Grand Tour

8. Jules goes to the airport and catches a plane. In disguise, Madame sits in the seat behind him. They land at a city that has sometimes been ruled by Poland, sometimes by Russia. Today it's the capital of an independent nation. It's about 250 miles (402 km) northeast of Warsaw, Poland, and about 500 miles (805 km) east and a little south of Copenhagen, Denmark.

9. Back at the hotel that evening, Jules and the Countess pretend that neither of them has left the city. They're off again the next morning for a resort on the Black Sea. The city lies about 275 miles (443 km) south of Kiev, its nation's capital. It's also a little more than 700 miles (1,127 km) southwest of Moscow, Russia, which *was* its capital until 1991. Who is that man Jules is talking to? He's not dressed for the beach! What is Jules handing him?

10. Jules tips his hand in the city where Europe meets Asia. It lies on the Bosporus, one of the straits connecting the Black Sea with the Mediterranean. It's also about 220 miles (354 km) northwest of its country's capital, Ankara. In the famous Blue Mosque, Jules passes stolen jewels to the man from the beach. The police close in. And where is Madame? Where, indeed?

Can you identify the cities and countries visited by Jules and Madame? Write their names on the lines below.

1. _____   6. _____
2. _____   7. _____
3. _____   8. _____
4. _____   9. _____
5. _____   10. _____

Desert Lands

"I've found out where they've hidden the gold," Agent X whispered urgently over the phone. "It's a city somewhere in the Middle East. Most of its people follow the Muslim religion."

"Well, gee, that only takes in all of North Africa and Southwest Asia," I said dryly.

"Wait, there's more. It's not near the sea. It lies at one end of a high, well-traveled mountain pass. It's a national capital. When the people pray, they face southwest toward Mecca."

The phone went dead. I took my atlas down from the bookcase to see if I could find the city Agent X was talking about.

1. I looked first at the region's largest city. It's also the largest city in Africa. It is more than 1,300 years old, but west across the Nile River are buildings that are 3,000 years older still! About 110 miles (177 km) northwest is Alexandria, this country's second-largest city. But there are no mountains around.

2. If you travel from there about 760 miles (1,223 km) southeast, you come to a city on the eastern shore of the Red Sea. The city's name is Arabic for "grandmother." There is an old legend that Eve, the "grandmother of us all," is buried here. For centuries, this city has been a port of entry for Muslim pilgrims going to Mecca, about 72 miles (45 km) southeast. Oops, wrong direction!

3. Continue about 745 miles (1,200 km) southeast, crossing the Red Sea again. You're on the Gulf of Aden, in the capital of a small country with the same name as the city. You're on the seacoast, however, and Mecca is to the northwest.

Desert Lands

4. About 765 miles (1,231 km) west and a little north of the last stop, you come to another national capital. Here the Blue and White branches of the Nile meet. The place is an ancient stopping point for camel caravans traveling between Egypt and central Africa. But there are no mountains nearby, and Mecca is northeast.

5. You travel northwest across the Sahara about 2,325 miles (3,741 km). You're in a city sometimes called "the White" because of its many white buildings. On a hill in the north end of town is the city's ancient *Casbah,* or fortress. The city is a capital, and there is a pass through the Atlas Mountains to the south. But the city is on the sea, and Mecca is to the east.

6. Take another long hop—eastward and a little south this time, about 2,335 miles (3,759 km). You're on the Tigris River, northwest of the Persian Gulf. About 1,200 years ago, this city was the capital of a great empire—and the setting for the famous story "Ali Baba and the Forty Thieves." It's a national capital today, but not the one I wanted.

7. How about the place that's around 500 miles (805 km) westward across the Syrian Desert? It's one of the oldest cities in the world, founded about 5,000 years ago. It lies along a trade route approximately 50 miles (80 km) inland from the Mediterranean coast. Trucks are the main trade vehicles there today, but outside the city, many people still live in tents and travel by camel.

8. I looked at the name of a city in the Judean Hills, a little more than 30 miles (48 km) from the Mediterranean coast at Tel Aviv. It's only about 135 miles (217 km) southwest of the last place I looked, but it's hard to get from one to the other. Their two countries have been officially at war for many years. This ancient city is a capital today, as it was 3,000 years ago. It is a holy city to Muslims, Christians, and Jews.

9. If you travel about 965 miles (1,553 km) east, you're in a land that was once called Persia. In the early 1600s, this city on the Zayandeh River was the capital of Shah (King) Abbas. He made the city a center for art. It's best known today for the beautiful rugs that are woven here. The city is a little more than 200 miles (322 km) south of its country's present capital, Teheran. Mecca is to the southwest, so at last I was looking in the right direction.

10. Aha! If you go northeast about 1,025 miles (1,650 km), you're in a national capital, on a river with the same name as the city. Its Muslim citizens face southwestward to pray. The city lies below the Hindu Kush mountains at the western end of the Khyber Pass, the ancient trade route to India. That was where the gold had to be! Silently I thanked Agent X for his useful information.

Can you identify the 10 cities and countries suggested by the clues? Write their names on the lines below.

1. _____   6. _____
2. _____   7. _____
3. _____   8. _____
4. _____   9. _____
5. _____   10. _____

African Safari

The Bates family was looking over the photographs from their trip to Africa. "We visited so many places they all run together in my mind," said six-year-old Jamilla Bates. "Maybe these pictures can help me sort them out."

1. One photo showed colorful murals painted on glass. "I remember this place," Mrs. Bates said. "It's the headquarters of the Organization of African Unity—'Africa Hall.' You remember that city. It's the capital of Africa's oldest independent nation, a little more than 600 miles (966 km) southeast of Khartoum, Sudan."

2. There was a picture of a herd of zebra. "I liked the wildlife best of all," 12-year-old David said. "I took that photo near Lake Victoria. We stayed in that hotel in the capital city about 185 miles (298 km) east of there and about 265 miles (426 km) northwest of Mombasa on the Indian Ocean."

3. The next picture showed an Arabian palace. "That was the Sultan's palace in the 1800s," Mr. Bates said with distaste. "He got rich trading in spices, ivory, and *slaves*."
   "But today that island city is part of an independent African nation," his wife reminded him. "It's a little less than 50 miles (80 km) north of the mainland capital city, Dar es Salaam."
   "What *is* the name of that city?" Jamilla asked. "It's the same as the name of the island. Begins with a 'Z,'" doesn't it?"

World Geography Mysteries
© The Learning Works, Inc.

African Safari

4. One set of pictures showed a colorful outdoor market. "We saw a lot of those in Africa," remarked David. "This one was in the capital of that large island nation about 250 miles (402 km) off the coast of Mozambique. I thought it was interesting that a lot of that country's early settlers came from Southeast Asia."

5. Another group of photos showed a modern city. "That's the richest city in Africa," Mrs. Bates said. "It's on top of a gold mine—literally! What a long struggle the people of that country had to win their freedom! The city is about 30 miles (48 km) south of the country's executive capital, and about 780 miles (1,255 km) northeast of Cape Town, the legislative capital."

6. One picture showed a wide, rundown street. "That's the capital of the country that changed its name, isn't it?" Jamilla asked.

   "That's right. The country was called Zaire when we visited, but it changed back to its old name in 1997," Mr. Bates said. "The country's capital is right across the Congo River from Brazzaville, the capital of another country with a similar name. It's confusing, but not to the people who live there."

7. "I remember that white church," Mrs. Bates said, looking at another picture. "It was in that capital just southeast of a lake with the same name as the country. I had a chance to practice my French there. It's near the borders of Cameroon and Nigeria."

8. There was a picture of the children playing on an ocean beach. A city of white buildings was in the background. "That city was a center for the slave trade, too," Mr. Bates remarked.

   "I know, Dad," David said. "It's also the capital of a country that took its modern name from an ancient African kingdom. You wouldn't let us forget that. The city is about 265 miles (426 km) east of Abidjan, Côte d'Ivoire, and about 265 miles (426 km) southwest of Lagos, Nigeria."

African Safari

9. There was a picture of a collection of mud-brick houses. They were of the same red-brown color as the surrounding desert land. "That was the town you said had such a glorious past, Dad," said David.

   "And so it did," said his father. "About 600 years ago, it was the center of a great empire. Do you remember its name—three letters, beginning with 'G'? It's on the Niger River, just under 600 miles (965 km) northeast of its nation's capital, Bamako."

10. The last photograph showed a street of red brick houses. The photo looked like it could have been taken in any American city of the 1800s. "That was the capital of the country that was founded by freed American slaves," Mrs. Bates said. "The city was named for the fifth president of the United States."

    "That's right," said Jamilla. "It's on the Atlantic coast, about 730 miles (1,175 km) southeast of Dakar, Senegal."

    The rest of the family looked at Jamilla in surprise. "See?" said her mother. "You *do* remember."

Can you identify the places the Bateses photographed in Africa? Write the names of the cities and their countries on the lines below.

1. _____   6. _____

2. _____   7. _____

3. _____   8. _____

4. _____   9. _____

5. _____   10. _____

World Geography Mysteries
© The Learning Works, Inc.

Half the World is Asia

Mr. Torres gave his class a quick quiz. Which continent, he asked, has more than one-fifth of the earth's land area? Which has nearly three-fifths of all the people? Which has both the highest and lowest points on earth? Which continent (other than North America) is closest to the United States? And which continent contains the only human construction visible from the orbiting Space Shuttle?

Many students knew that the answer to all five questions was the same—Asia. Did you? Let's see what else you can find out about the map of Asia.

1. Let's start in the capital of the world's most populous nation. The city has been a center of government off and on for more than 2,000 years. It lies about 90 miles (145 km) from the sea, at Bo Hai Gulf. It's only about 30 miles (48 km) south of the Great Wall of China—which astronauts can see from Earth's orbit.

2. Now let's head about 1,180 miles (1,900 km) south and a little west. We're in the tropics now. We're in a city near the mouth of the Zhu (Pearl) River, about 80 miles (129 km) northwest of Hong Kong. This city is China's great international trading center. Most of the Chinese people who emigrated to the United States before 1949 came from the countryside around here.

3. Let's make one more stop in China. This time let's go inland, about 600 miles (966 km) northwest of the last place we stopped. We're in a port city on the Yangtse River (or Chang Jiang), about 900 miles (1,448 km) west and a little south of China's largest city. The beautiful gorges on the Yangtse nearby have been painted through the centuries by many of China's greatest artists.

Half the World is Asia

4. Let's go to the capital of another country. It's about 1,250 miles (2,013 km) north of our last stop, and about 465 miles (748 km) east of the country's second-largest city, Uliastay. Outside of these two cities, many of this country's people still follow a traditional way of life, herding sheep, goats, horses, and camels.

5. Now let's travel to yet another country. We go about 325 miles (523 km) northwest. We're in a city on a river that feeds into Lake Baikal. The country we're in lies partly in Asia and partly in Europe. The city is closer to the capitals of China, Japan, and India than it is to its own capital, which lies more than 2,600 miles (4,186 km) west and a little north of here.

6. About 1,425 miles (2,293 km) southeast of our last stop, we come to a Pacific Ocean port. We're still in the same country, but near the borders of China and Korea. This city is an important naval base and trading port. A train called the Trans-Siberian Express travels between here and the country's capital. The journey takes a week! The name of the city has 11 letters, beginning with "V."

7. We travel about 465 miles (748 km) southwestward to another national capital. It lies on a peninsula, between the Sea of Japan and the Yellow Sea, about 590 miles (950 km) southeast of the first place we visited on this tour. The city hosted the 1988 Summer Olympic Games. We pronounce its name like an English word that means "only."

Half the World is Asia

8. Now let's go about 515 miles (830 km) southeast to a busy industrial center. We're in a different country again, in its second-largest city. It's about 250 miles (402 km) southwest of the capital, which is the most populous city in the world. An efficient way of traveling between the two cities is the Bullet Train.

9. We travel about 650 miles (1,046 km) northeast. We're still in the same country, but on a different island, Hokkaido. We're in the island's capital and largest city. It's a favorite destination for vacationers because of the nearby ski slopes and hot springs.

10. Finally, just to get an idea of the size of Asia, let's travel about 3,570 miles (5,748 km) west and a little south. We're in an ancient caravan city which is also the capital of a new nation. It's about 930 miles (1,496 km) west and a little south of Urumqi, China, but only about half that distance north of Kabul, Afghanistan.

Can you identify the 10 places we visited on this tour? Write the names of the cities and countries on the lines below.

1. _____   6. _____
2. _____   7. _____
3. _____   8. _____
4. _____   9. _____
5. _____   10. _____

Half of Asia is Southern Asia

"They really thought they had me," said ace reporter Wally "Whiz" Worldbeater at his press conference. "They wanted me to name my sources on the Kangalia incident. They had me tied up in the back of their plane under a sheet of canvas. They kept flying from place to place, trying to confuse me. Every now and then, someone would come back and threaten to throw me out without a parachute if I didn't talk. But they didn't know about my trusty hand-held navigational computer. I knew where I was at all times.

1. "They grabbed me in one of the most beautiful cities in the world. It's in the Vale of Kashmir. They seized me in a tea shop beside the Jhelum River. What? What country? Well, it's in India now, but Pakistan claims it. Let's just say I was about 400 miles (644 km) northwest of the capital of one, and about 100 miles (161 km) northeast of the capital of the other."

2. "There was no question where they took me, though. They flew about 1,050 miles (1,691 km) south and a little west. I could smell the sea—the Arabian Sea, no doubt. I had to be in the largest city in India, about 720 miles (1,160 km) southwest of the capital."

3. "When we took off again, we headed northeast. We flew almost 1,180 miles (1,900 km) this time. We must have passed just north of Calcutta. We landed in the capital of the country that borders India on the east. You remember; it was East Pakistan until 1971. It was hot and humid—monsoon season, of course. At least we were spared the flooding that often devastates that country."

4. "I was hardly aware of our next landing. The city is so high above sea level, it felt like we were still in the air. It's a national capital, in the foothills of the Himalaya, the world's highest mountain range. We were about 420 miles (676 km) northwest of the last stopping place and only about 100 miles (161 km) southwest of Mount Everest."

5. "Our next hop took us about 780 miles (1,255 km) southeast. When I heard temple bells, I knew where we were. There's a famous poem about the road to this place. It's on the Irrawaddy River, about 360 miles (580 km) north of the country's capital, Rangoon."

6. "About 635 miles (1,022 km) further southeast, I could smell the sea again. This time it had to be the Gulf of Thailand. I was in a city just a few kilometers north of there. You know the one I'm talking about. It's a metropolis that's grown about four times in size since my dad was here during the Vietnam War. A lot of the city's famous canals have been filled in for streets and freeways."

7. "Speaking of Vietnam, I'm sure that's where we went next. The city where we landed was about 465 miles (748 km) southeast of our last stop. A lot of places get their names changed after a war, and this city was once called Saigon. It's about 700 miles (1,127 km) south and a little east of Hanoi.

Half of Asia is Southern Asia

8. "Soon we were off again. Judging by the air currents, we were over the open sea for most of the flight. We flew about 685 miles (1,102 km) southwest to a city at the tip of the Malay Peninsula. A city-state, I should call it—the city makes up most of the country. It's a small nation but a wealthy one."

9. "Soon we were off again, flying northeastward over the South China Sea. We flew about 1,485 miles (2,390 km), so I knew where we had to be. It's the capital of an island nation, named for a former king of Spain. The country consists of more than 7,000 islands. The city has given its name to a color used for large envelopes and folders. What's that? Stick to the point? You've got a lot of nerve after what I've been through!"

10. "We flew southwest, about 1,735 miles (2,793 km). We flew over the island of Borneo and landed in the capital of another island nation. It's on an island whose name is slang for 'coffee,' and for good reason. I know that area well. I lived for a year on Bali, the next island to the east. I have friends there. I made a daring escape and found my way into the city. Anymore questions?"

Can you identify the 10 cities and their countries suggested by the clues? Write their names on the lines below.

1. _____ 6. _____
2. _____ 7. _____
3. _____ 8. _____
4. _____ 9. _____
5. _____ 10. _____

World Geography Mysteries
© The Learning Works, Inc.

Lands Down Under

Australia is many things. It's the world's smallest continent and the world's largest island. Often you hear Australia called "the land down under," and it is—on a globe, anyway! Of course, there are places on the globe that are even further "down under" than Australia. Let's explore some of the "lands down under."

1. Let's start where Australia began as a country about 200 years ago. Of course, people have lived there for at least 40,000 years. But in 1788, England started a prison colony on Australia's Pacific coast. (Previously, they had sent convicts to America, but the Revolutionary War put a stop to that!) The settlement near the shore of Botany Bay grew into Australia's largest city. It's about 155 miles (250 km) northeast of Canberra, the Australian capital.

2. Australia's second-largest city faces south onto Port Phillip Bay. It's about 285 miles (459 km) southwest of Canberra. The city was founded as a settlement of sheep farmers in 1835. Sheep are still important to Australia's economy, but not in this city. It has grown into Australia's leading business center.

3. Australia's western coast is cut off from the rest of the country by a large desert. Its only large city lies about 1,925 miles (3,100 km) west and a little north of Canberra. The city started as a mining camp and has become a state capital. It became famous in 1962 when the city turned on its lights as a beacon to orbiting American astronaut John Glenn.

4. Almost in the center of Australia, about 1,215 miles (1,956 km) northwest of Canberra, stands the only city for miles around. It's in the middle of a desert, and its name suggests that it has a source of water. It lies just south of the Tropic of Capricorn. Visitors to Australia are often surprised that they don't see kangaroos hopping down city streets. Here is one place where it happens!

5. Now let's leave Australia for some of the outlying islands. North of the continent is a large island. The people who live here are called Melanesians. The western half of the island is part of Indonesia, but the east is an independent country. Its only large city lies on a peninsula. It's about 1,300 miles (2,093 km) northwest of Brisbane, Australia, and about 280 miles (451 km) southeast of Mount Wilhelm, the island's highest point.

6. Another Melanesian island capital is found about 2,080 miles (3,350 km) east of Townsville, Australia, and about 2,175 miles (3,500 km) northeast of Canberra. Many people came to these islands from India as workers during the 1800s.

7. Southeast of Australia is another English-speaking nation. It is made up of two large islands and many smaller ones. The largest city is about 310 miles (500 km) north of the capital, Wellington, and about 1,430 miles (2,302 km) east and a little south of Canberra, Australia. The first settlers of these islands were the Maoris, a Polynesian people. They came here from other Pacific islands more than 1,000 years ago.

Lands Down Under

8. The Polynesians originally came from Southeast Asia. They inhabited hundreds of islands, from Madagascar to Hawaii. One Polynesian island capital is found about 2,035 miles (3,276 km) northeast of Brisbane, Australia, and about 560 miles (901 km) southwest of the islands of Samoa.

9. A third people of the Pacific islands are the Micronesians. One Micronesian island is a United States dependency. Its capital is home to a United States Air Force base and an international airport. It's in the Mariana Islands, about 3,540 miles (5,700 km) north of Melbourne, Australia, and about 3,790 miles (6,102 km) west and a little south of Honolulu, Hawaii.

10. Finally, let's head south to earth's least-populous continent. There are no cities here, so we're looking for a mountain. It's 12,280 feet (3743 m) tall. It lies near the edge of the Ross Sea, about 2,515 miles (4,050 km) south and a little west of Wellington, New Zealand. Oh, and it's only about 850 miles (1,368 km) from the South Pole—north, naturally.

Can you identify these 10 places? Write the names of the cities and countries on the lines below. For number 10, write the name of the mountain and continent.

1. _____   6. _____
2. _____   7. _____
3. _____   8. _____
4. _____   9. _____
5. _____   10. _____

Deeper Into the World

Mountains and Deserts

"I'm tired of hearing you boast about all the mountains you've climbed!" Debra Sands said with irritation. She and Sir Patrick Peaks were sitting in the lounge of the Explorers' Club. "What's a mountain but a big pile of rocks anyway? Now, a desert! That's—"

"Boring, boring, boring!" Sir Patrick interrupted. "You've seen one desert, you've seen them all."

"That shows how little *you* know," Debra shot back. "I'll make a wager with you. I'll describe five deserts I've explored. You describe five mountain ranges. I'll bet I can identify more of yours than you can of mine. In fact, I'll bet you dinner on it."

"You're on!" said Sir Patrick.

1. "I'll start with an easy one," Debra said. "I once trekked across the largest desert in Asia. I started from Altay at the foot of the Hangayn Nuruu Mountains and traveled by camel caravan to Choybalsan. And since you think all deserts are the same, let me tell you, it was cold! And I hardly saw a sand dune anywhere!"

2. "I'll give you a break, too," Sir Patrick said. "I once hiked down the spine of the mountain range that separates Europe from Asia. From north to south, that's about 1,640 miles (2,639 km). It took me 90 days, not counting the week I spent resting in Sverdlovsk."

3. "Once I hiked across the driest desert in the world," said Debra. "In parts of it, rainfall has never been recorded! The clouds don't drop their moisture until they reach the east side of the Andes. This desert is rich in minerals, and there are mining towns here and there. Still, when I reached Antofagasta, I spent a lot of time gazing at the Pacific Ocean. It was good to see so much water!"

4. "I do know a thing or two about deserts," Sir Patrick rejoined. "Once I scaled the mountain range that overlooks the Sahara from the northwest. I climbed its highest peak, Jebel Toubkal, and followed the ridge to Irhil m'Goun. I suppose I was ready for a glass of water by the time I got back to Marrakech."

5. "You've been to Africa, then," Debra said. "I've spent time in the Sahara, but there's another African desert that I prefer. I got to know it well when I lived with a San family in Botswana. We hunted with poison darts and drank out of shallow pools. A town like Ghanzi or Kang was the closest we came to city luxuries."

6. "So you have known privation." Sir Patrick looked at Debra with respect. "I'll tell you about a mountain range I've never crossed, except in the passes. But I have climbed its highest peak, Tirich Mir—25,230 feet (7,690 meters). The range is in central Asia, and it divides the waters of the Indus River valley from those of the Amu Darya."

7. "There's a desert I'll take you to if you ever visit Tucson," Debra said. "It ranges south toward Hermosillo, Mexico. It has everything you think a desert should have—sandstorms, tarantulas, rattlesnakes. It even has saguaros—you know, the cactuses with the arms. It's the only place in the world where they grow."

Mountains and Deserts

8. "I'll take *you* to my favorite mountains—if you don't scare easily. Now, don't look at me as if I were mad!" Sir Patrick leaned forward. "It's a small range in the middle of Germany, between Goslar and Eisleben. They're mere hills, really. But it's where the brothers Grimm gathered many of their famous fairy tales—the actual home of 'Hansel and Gretel!'"

9. "There's one desert I've studied but never actually crossed," Debra said. "I couldn't find a guide. Even the Arab Bedouins avoid it. It's in a corner of the Arabian Peninsula. The United Arab Emirates is to the north, and Oman is to the east and south. Its name in Arabic means 'the empty quarter.'"

10. "I have one last chance to win," said Sir Patrick. "Name the longest mountain range in the world. It runs from the Arctic to the Antarctic, but few of its peaks have ever been climbed. Some of them are volcanic, but that's not why. I've climbed Hekla, east of Reykjavik, and looked down into its crater. And I've flown over the southern end of the range, near Tristan da Cunha. . . ."

"You won," Debra said later, in the restaurant. "But I still say that last one was a trick question."

Can you identify the deserts and mountain ranges suggested by the clues? Write their names on the lines below.

1. _____   6. _____
2. _____   7. _____
3. _____   8. _____
4. _____   9. _____
5. _____   10. _____

Lakes and Inland Seas

Somewhere on earth is a sea that is not a sea. It is salt water, but it lies inland from the ocean. It is nearly impossible to drown in it, but no fish live in it. It is on the border between two countries. Where and what is it? Let's see if we can find out by taking a closer look at some of the world's lakes and inland seas.

1. Let's start with the world's largest inland body of water. To the north is Europe. To the south is Asia. It is also salt water and it *does* hold fish. Only two countries bordered it before 1991, but today there are five.

2. Let's go north, to the Arctic Circle. Here is a lake named for an animal. It may be hard to drown in it, but only because it is too cold for swimming. It is fresh water, though, and it lies entirely within one country—Canada.

3. Now let's try an east Asian sea. Islands surround it on all sides—Kyushu, Shikoku, Honshu, and many smaller islands—but they are all part of the same country. It is a true sea and is part of the Pacific Ocean. Two straits connect it to the Philippine Sea.

4. You sail into Europe's largest inland sea through the Skagerrak and the Kattegat. No fewer than five national capitals line its shores—uh-oh, that's too many! It is connected to the Atlantic. It is well fished, too. Do you like herring?

5. Here is a lake, high in the mountains of South America. It is bordered by two countries—so far, so good. It is entirely inland. It has no outlet to the sea, though it is drained by a river into Lake Poopó. It is fresh water, however, and it's full of large trout.

6. Let's try one of Africa's great lakes. It is the longest lake in the world and the second deepest. Uh-oh, sorry—make that the longest *fresh-water* lake. And no fewer than four countries share its shores, from Burundi in the north to Zambia in the south.

Lakes and Inland Seas

7. How about a North American lake? It lies on an international border. A large city is near its western shore. It's part of a waterway connecting lakes Huron and Erie, but it's fresh water.

8. Here's a promising one. It's a salt-water lake—when it exists. Sometimes it completely dries up. That makes it hard to drown in and not good for fish. Wait—this lake lies entirely within Australia. It's about 400 miles (644 km) northwest of Adelaide.

9. Now let's check out a "sea" in central Asia. It's salt water. It's bordered by two countries. But where did it go? Once this was the world's fourth-largest inland body of water, but not any more. Irrigation has drained off the two main rivers that feed it, the Amu Darya and Syr Darya. No fish live in it—it's too polluted.

10. That last place was depressing—a dead sea. . . . Wait a minute! What about that little lake in southwest Asia? It's between two countries, Israel and Jordan. It is about nine times as salty as the ocean. That's too salty for fish. You can float in it easily. Just try sinking in it—bet you can't!

Can you identify these 10 bodies of water? Write their names on the lines below.

1. _____ 6. _____

2. _____ 7. _____

3. _____ 8. _____

4. _____ 9. _____

5. _____ 10. _____

Around the World, Island to Island

I hadn't meant to sail around the world. I was only trying to find the artist Carol Coral to interview her for my magazine. Rumor had it that she had gone to live on an island to get away from it all. That would have left me about 40,000 places to look—if I hadn't had a list of the most likely places she would have gone.

1. I cast off from an island off the east coast of the United States. Its name was once New York Island, but since the 1830s it's been known by a different name. Its southern end is on a bay, but its other three sides are bordered by rivers—the Hudson, the Harlem, and the East.

2. I set a course southeastward. It was about 670 miles (1,078 km) to my first port of call. It was a group of islands, a dependency of Great Britain. It's a popular destination for beach vacations, but these are not tropical islands. They lie due east of Savannah, Georgia. They have a warm climate because of the Gulf Stream.

3. From there, I sailed east across the Atlantic, past Gibraltar and into the Mediterranean Sea. I landed at an island about 60 miles (97 km) south of Sicily. It's the main island of a group of islands that form an independent country. The island and the country have the same name. The country's language is an interesting blend of Italian and Arabic. Because it used to be part of the British Empire, many people here speak English, too.

4. I sailed on southeastward to Egypt. I passed through the Suez Canal, into the Red Sea, and out across the Arabian Sea. At last I made landfall at a large island that's about 2,000 miles (3,220 km) northeast of Mogadishu, Somalia, and just off the southeast coast of India. Many groups of people from different parts of the world have settled and ruled here in times past. Today, it's an independent country whose capital is Sri Jayewardenepura.

5. I set a course east by southeast for one of the largest islands in Indonesia. As a matter of fact, it's the sixth largest island in the world. It lies diagonally astride the equator, just west of the Malay Peninsula. I called at the port of Padang, one degree south of the equator.

6. I set sail again, passing south of the Sunda Islands and north of Australia. Soon I was in the Pacific Ocean, bound for an island whose name is synonymous with romance and adventure. It's one of the Society Islands. Polynesian sailors used to travel by canoe between here and Hawaii, about 2,500 miles (4,025 km) north and a little west. I put in at Papeete, the island's capital. The French painter Paul Gauguin worked here once—why not Carol Coral?

7. Just south of the Tropic of Capricorn, about 1,350 miles (2,174 km) southeast of my last port of call, is a tiny island. Only about 60 people live here. Adamstown is the only settlement. In 1789, sailors on the British naval ship *Bounty* mutinied against their captain, William Bligh. They set him adrift in a small boat and sailed the *Bounty* to this island, where they hoped no one would ever find them. Their descendants still live on the island.

Around the World, Island to Island

8. I sailed southeast, heading toward a large island just north of stormy Cape Horn. The island is divided between two countries, Chile and Argentina. Its name is Spanish for "Land of Fire." I sailed through the Strait of Magellan and back into the Atlantic.

9. About 350 miles (563 km) east of the strait, I paused for rest at a group of islands. They're a British dependency, and their capital is Stanley. If you started from Louisbourg, the easternmost city in Canada, and sailed a *long* way due south, you'd reach these islands. Britain and Argentina fought a war over these islands in 1982. The Argentines call them the Malvinas.

10. I found Carol more than 3,500 miles (5,635 km) northeast of my last stop. I almost missed this tiny island. A famous prisoner was once held captive there, as Carol well knew. "Napoleon couldn't escape from here," she said. "How did you find me? Ascension Island, the nearest land, is almost 700 miles (1,127 km) northwest."

    "Pinpoint navigation," I said modestly.

Can you identify the islands and island groups suggested by the clues? Write their names on the lines below.

1. _____   6. _____
2. _____   7. _____
3. _____   8. _____
4. _____   9. _____
5. _____   10. _____

New on the Map

Incredible news, Mike—the time machine works! It's 1998 where I am. You won't believe some of the changes! But before I tell you about personal computers, VCRs, and cable TV, let me tell you about the map. There's all these new countries that don't exist in 1970. I've got to find some books and catch up on what's been happening, but let me give you a few examples.

1. "First of all, the Soviet Union split up in 1991. Russia's still the biggest country in the world, but it doesn't have that empire any more. There's a large new country southwest of Russia. Actually, it's a very old country. At different times, it was called 'Ruthenia,' and 'Little Russia.' Its people have their own language and their own culture. Now they have their own country. Some of its neighbors include Belarus, Poland, and Moldova."

2. "There's another old country that's risen again. I mean, this place was an independent kingdom in, something like, 95 B.C., but it's been ruled by one empire or another for most of the time since then. Now it's free again—can you believe it? It borders Turkey on the northeast, and its capital is Yerevan."

3. "The Soviet Union isn't the only country that has broken up. What used to be Yugoslavia is now five different countries—and from what I hear, they're still fighting over the pieces. There's a country called Slovenia in the north, and one called Macedonia in the south, and three others in between. This one has Zagreb as its capital. It projects outward like two arms west and north of Bosnia and Herzegovina."

4. "Then there's that country that used to be north of Austria and Hungary, and south of Poland. It's been two countries since 1993. Prague is the capital of something called the Czech Republic, but there's a new country east of it now. Its capital is Bratislava."

5. "Most of the big changes have happened in Europe, but there are new countries all over the globe. Take 'British Honduras,' for example. It isn't there any more, and neither are the British. It changed its name in 1973 and became independent in 1981. It doesn't even border on regular Honduras, but it does border on Mexico and Guatemala."

6. "The British seem to have lost most of their empire. All those islands in the West Indies that used to be British colonies—they're independent now. Take the group of islands that's about 200 miles southeast of Puerto Rico. It's been a free country since 1983, with its capital at Basseterre. Just to the east is Antigua and Barbuda, which is another independent country."

7. "Just about all the old European colonies in the Americas are gone. You know 'Dutch Guiana,' in South America? It isn't Dutch any more. It's an independent country, and its capital is Paramaribo. To the west is Guyana, which used to be British Guiana. To the south is still good old Brazil."

8. "There are a few more countries in Africa now too. That territory we know as South West Africa, between South Africa and Angola—it's been independent since 1990. It has a new name, too, but Windhoek is still the capital."

New on the Map

9. "Then there's Ethiopia—it's smaller than it used to be. Apparently they had a civil war. The northern part broke away to form its own country in 1993. Its capital is Asmara."

10. "In Asia, too, new countries have broken away from the old Soviet Union—places that had been ruled by Russia for centuries. There's Uzbekistan, Tajikistan, and one that's really out in the middle of nowhere. Actually, it's in the middle of Asia. It's got Kazakstan to the north and China to the east and south. Its capital is Bishkek."

"Well, I've got to end this report. There's something I want to watch on MTV, and then I'm heading over to an arcade to play video games. What? Oh, right. I'll explain that in my next report."

Can you identify these 10 independent countries that didn't exist in 1970? Write their names on the lines below.

1. _____   6. _____
2. _____   7. _____
3. _____   8. _____
4. _____   9. _____
5. _____   10. _____

Follow the River

"We've intercepted this coded message on the Internet, Agent Intrepid," Agent Y said. "We think it was posted by Ivory Hunter, the notorious killer of endangered species, for one of her buyers."

"'Meet me on the river with the cash,'" I read. "'You go downstream, I'll go up, and we'll meet someplace in between.' *Which* river?" I asked Y.

"The message doesn't say," Y replied, producing another scrap of paper. "But we have reason to believe it's one of these."

"But this list names rivers all over the world!" I said.

"Then you'd better send out your people at once," Y replied.

1. I sent Agent A to the river where civilization began more than 5,000 years ago. His knowledge of Arabic would come in handy there. The river rises in the Pontic Mountains of northeastern Turkey and ends in the marshes at the head of the Persian Gulf. It flows past the ruins of many ancient cities, including fabled Babylon.

2. Agent B went to Africa, where Ivory had done most of her dirty work. B patrolled a river that begins in Guinea and ends in the Gulf of Guinea. It flows past farmland, desert, and swamp, through or beside five countries. Two of these countries have been named for the river.

3. I sent Agent C to a colder climate. He explored a river where people still make a living trapping furs—but not of endangered species! He started at the Great Slave Lake and followed the river northwest to its mouth at the Beaufort Sea. That's an arm of the Arctic Ocean. I hope C remembered to pack his long underwear.

Follow the River

4. While others braved deserts or ice, Agent D was enjoying the cafes of Vienna and Budapest. Her river starts in the Black Forest of Germany. It flows through four national capitals on its way to the Black Sea. I'm sure when this is over she'll bore us all with stories of how "beautiful" and "blue" it is.

5. Agent E speaks no language but English. I sent him to a real "sweetheart" of a river. It rises in the Great Dividing Range and flows mainly southwestward. It's the longest river on its continent, but it does not empty directly into the sea. It ends at the Murray River, which in turn flows into the Indian Ocean.

6. Agent F is a tall, powerful woman. I hope she appreciated where I sent her. Her river starts in the Andes Mountains of Peru but flows mainly through a neighboring country. It ends up in the Atlantic, north of Marajó Island. It's a *huge* river, holding one-fifth of all the fresh water on earth. I thought Ivory might turn up in the city of Manaus. It's near the rain forest and there are a lot of rare animals nearby.

7. "I like waterfalls," Agent G once told me. I hope he enjoyed seeing mighty Victoria Falls. It lies far up the river where I sent him, and you can't get there by boat. The borders of four countries meet near the falls. One of them, which lies north of the river for part of its course, is named for the river.

Follow the River

8. Agent H went to Asia—and good luck to her. She had to follow a river that Hindu people consider sacred. It rises in the Himalaya Mountains and enters the sea at the Bay of Bengal. At holy cities such as Varanasi and Allahabad, thousands of pilgrims come to bathe in its waters. It also flows through Calcutta, one of the world's largest cities. If Ivory was there, H would have to find her in a crowd.

9. I sent Agent I to patrol the longest river in Europe. It lies entirely within one country. Its source is about 200 miles (322 km) southeast of St. Petersburg, and it ends up in the Caspian Sea. The people of this country sing a sad song about slaves that used to push boats along this river.

10. I caught Ivory Hunter myself, and I didn't have to leave home to do it. I found her trying to peddle a load of rhinoceros horn in Dubuque, Iowa—on the longest river in the United States. She's in jail now. I hope my operatives enjoyed their travels.

Can you identify the 10 rivers suggested by the clues? Write their names on the lines below.

1. _____ 6. _____

2. _____ 7. _____

3. _____ 8. _____

4. _____ 9. _____

5. _____ 10. _____

Dead Reckoning

"So you think you know the world?" the mysterious old man said.

"I've traveled quite a bit, yes," I said, puzzled. He clearly was a gentleman of means, though when he'd walked into the cafe I'd taken him for a poet or scholar.

"Then this may be your lucky day," he said. He produced a scrap of paper from his dusty jacket pocket. "I have a list of 10 places. Each is identified only by two lines of verse and its position by latitude and longitude. Can you tell me what each of them is? If so, I will pay your way for a trip around the world."

Well, friends, I met the challenge. I'm off tomorrow for the journey of a lifetime. Let's see how well you can do with the same list. Unlike me, you may use a globe or an atlas.

1. A city famous for its art
   Seems like the perfect place to start.
   43–44°N, 11–12°E

2. A high mountain is your next stop:
   Hope you're not there when it blows its top!
   19°N, 98–99°W

3. An island with a repetitious name—
   Enjoy its beaches if you win this game.
   14–15°S, 170–171°W

4. Your next stop is a path between the seas;
   To pass through its locks, you won't need any keys.
   9°N, 79–80°W

5. This inland seaport is a capital city;
   To not know its name would be a pity
   25–26°S, 57–58°W

6. An ancient city, once famous for its might!
   Northeast of a modern capital is the site.
   36–37°N, 10–11°E

Dead Reckoning

7. An exciting place, though you might not find it pretty:
   In the world's most populous country, its largest city.
   31-32°N, 121-122°E

8. A princely family rules this tiny state.
   Visit in winter—I hear the skiing is great!
   47-48°N, 9-10°E

9. Lion, elephant and zebra roam these grasses—
   More wildlife than in any science classes!
   2-3°S, 34-35°E

10. Climb another sleeping volcano:
    People call it sacred, so please don't say no.
    35-36° N, 138-139° E

Can you identify all these places? Write the names of the cities, geographical features, and countries on the lines below.

1. _____   6. _____

2. _____   7. _____

3. _____   8. _____

4. _____   9. _____

5. _____   10. _____

World Geography Mysteries
© The Learning Works, Inc.

The Complete World Traveler

A Capital Adventure

I was on the trail of Connie Soeur. She was trying to sell a painting by the great French artist Matisse. The only trouble was, it wasn't really by Matisse. It was a clever forgery that Connie had painted herself. My name is Art Copp. My job was to follow Connie from world capital to world capital until I saw the phony painting change hands.

1. Connie's home town has been the capital of the art world since about 1940. It's the capital of a lot of things—business, book publishing, the American theater, and sheer American bigness. It's not the capital of the United States, though. It's about 200 miles (322 km) northeast of that place, and just across the Hudson River from Jersey City, New Jersey.

2. Connie knew better than to try to peddle her fake in a city where so many people knew her. I trailed her "across the pond," to the largest city in Europe. Now, *here* was a capital. It had once been the capital of a world empire. It lies on the River Thames, about 330 miles (531 km) southeast of Edinburgh, Scotland, and about 160 miles (257 km) north of Caen, France. Would she be so bold as to try to sell her painting to the Queen?

3. She still had the painting when she left on the night train. I was in the next car as we sped through France and across the Alps to Milan. About 300 miles (483 km) farther southeast, she got off in another city that had been the capital of an empire. And what an empire! About 2,000 years ago, it held all or part of more than 30 of today's countries! Today it's still a national capital, an art capital, and the capital of the world's largest religious group. It's about 120 miles (193 km) northwest of Naples.

A Capital Adventure

4. I trailed Connie to the airport. She flew to a city on the Gulf of Finland. It used to be called Leningrad, but a few years ago it took back its old, historic name. It's the second-largest city in Russia, about 400 miles (644 km) northwest of the capital. However, many Russians think of this city as their real capital.

5. Either people were spotting the fake, or Connie's price was too high. She left Europe, still carrying the painting. She flew to a national capital, Riyadh, where there was plenty of money still around from the oil boom of the 1970s. It's about 500 miles (805 km) northeast of an even more important "capital" in the same country—the religious center for more than one billion people.

6. Connie's next stop was about 1,700 miles (2,737 km) southeast of Riyadh, in the largest city in southern Asia. Part of it is on an island, about 1,030 miles (1,658 km) southwest of this country's second-largest city. Neither of them is its nation's capital, but this great port city is its business and industrial center. More movies are produced here than in Hollywood. I spotted Connie having dinner with a movie star, but he didn't buy the painting.

7. Where would Connie go next? Where else but to the largest city in the world. It's on Asia's Pacific coast, about 165 miles (266 km) northeast of Nagoya. The port of Yokohama is part of its metropolitan area. European art is popular here, but no one was buying Connie's painting.

8. Connie was leading me around the world. We crossed the Pacific to the largest city in South America. It's not its nation's capital, but it is its center for business and industry. It's about 225 miles (362 km) southwest of South America's second-largest city, which also is not a capital.

A Capital Adventure

9. Connie still had the painting when she boarded a plane northward to the world's second-largest metropolis. It's about 685 miles (1,102 km) south of San Antonio, Texas, and about 1,280 miles (2,060 km) southwest of Miami, Florida. In a museum, I overheard her offer to sell the painting to a dealer. Connie spoke fluent Spanish.

10. I followed Connie back to the States. I nabbed her in our second-largest city, the entertainment capital of the world. It's about 115 miles (184 km) northwest of San Diego, California, and about 350 miles (563 km) southeast of San Francisco. Connie offered me an "original" cartoon cel of Mickey Mouse if I let her go. I suspected that she, not Walt Disney, had painted it.

Can you identify the cities suggested by the clues? Write the names of the cities and their countries on the lines below.

1. _____   6. _____
2. _____   7. _____
3. _____   8. _____
4. _____   9. _____
5. _____   10. _____

Don't Forget Your Camera!

Good evening, and welcome to "Chance of a Lifetime." It's the game show that gives you a chance to win fabulous prizes! Tonight, we're offering a trip around the world. You'll visit 10 of the most popular tourist destinations on the globe. To win, all you have to do is be the first to identify the sites or nearby cities. Are you ready? Let's go!

1. The first stop on your tour will be in Mexico! This great Pyramid of the Sun stands at an ancient city about 25 miles (40 km) northeast of Mexico City. You can climb the pyramid yourself to the temple at the top—but first you'll have to name the city.

2. Your next stop is Blarney Castle in Ireland. Come and kiss the Blarney Stone, set in the castle in the year 1446. According to legend, a person who kisses the stone is given the gift of "blarney"—the power to charm and persuade people with speech. You'll stay in a city in southwestern Ireland. It's just southeast of Blarney, and about 135 miles (217 km) southwest of Dublin.

3. From there, it's a short hop to England. This ancient stone circle was built around 4,000 years ago. It's Stonehenge, of course. It's near a city that's been around nearly as long as Stonehenge itself. You'll find it just a few miles east of Wilton, and about 80 miles (129 km) southwest of the center of London.

4. You won't need to rent a car when you visit this city in Italy. It's built on about 120 islands connected by bridges. The streets are under water—they're all canals. People get around by boat. The city is at the head of the Adriatic Sea, about 65 miles (105 km) east of Verona.

5. Your next stop will be Egypt. You'll see the pyramids, but this great temple is at a city about 310 miles (500 km) southeast of Giza, where the pyramids are located. It's on the east bank of the Nile, about 110 miles (177 km) northwest of Aswan.

6. This abandoned city in Africa will be your next stop. It was built by the Shona people about 1,000 years ago. Its name in Shona means "house of stone." It's the same as the name of the modern country it's in today. You'll find it about 170 miles (276 km) south and a little west of the capital, Harare, and less than 20 miles (32 km) southeast of the town of Masvingo.

7. This is beautiful Angkor Wat in Cambodia. Nearly a square mile in size, it was a temple, an astronomical observatory, and the tomb of the king who built it. It lay near the city of Angkor Thom, home to a million people in the 1100s. For hundreds of years it lay abandoned in the forest. Today it stands restored, just northwest of a small city whose two-word name begins with an "S" and an "R." It's about 50 miles (80 km) northeast of Battambang, and about 150 miles (242 km) northwest of Phnom Penh.

8. Here is another sight everyone knows—the Great Wall of China. But where in China was this picture taken? The wall is more than 1,500 miles long! It's a small city right beside the wall, about 100 miles (161 km) northwest of Beijing, and about 235 miles (378 km) northeast of Taiyuan.

9. This beautiful mountain is a national symbol of Japan. It's 12,388 feet high (3,776 m), but more than 50,000 people climb it every year. It's about 60 miles (97 km) southwest of Tokyo, and a little more than 20 miles (32 km) southeast of Kofu.

10. Your last stop will be this tourist landmark in Florida. It's the home of Walt Disney World and many other theme parks. You'll find it about 80 miles (129 km) northeast of Tampa, and about 125 miles (201 km) southeast of Jacksonville. Say hello to Snow White and Pinocchio for us if you're one of the lucky winners!

Can you identify the cities and sites suggested by the clues? Write their names on the lines below.

1. _____   6. _____
2. _____   7. _____
3. _____   8. _____
4. _____   9. _____
5. _____   10. _____

For Those Who Love Adventure

She is Carolina Smith, archaeologist and adventurer. He is Dr. Mauvais, villain and scoundrel. He has stolen a treasure—the Lamp of Aladdin—from a cave high in the mountains of western China. It can give Mauvais powers beyond imagination, unless Carolina stops him. She is prepared to follow him to some of the most out-of-the-way places on Earth. Can she keep him from seizing the planet as his own plaything?

1. Mauvais has a hideout in an ice cave on the side of K2, the world's second-tallest mountain. Sometimes it's called Dapsang, or Mount Godwin Austen. It's in the Karakoram range, about 125 miles (78 km) east of a city in Pakistan. The city's name has six letters, beginning with "G." Carolina finds the cave deserted.

2. On a hunch, Carolina goes to Mount Kinabalu. It's a sacred mountain on an island on the equator. The island is divided among Malaysia, Indonesia, and the small state of Brunei. Mauvais has been spotted in a city on the northern coast of the island. It's about 35–40 miles (56–64 km) southwest of the mountain, and about 105 miles (169 km) northeast of the capital of Brunei.

3. In a tea shop, Mauvais has scribbled a line of poetry on a napkin. It reads, "A rose-red city half as old as time." To Carolina, this can mean only one place. It's an ancient trading city in present-day Jordan, about 115 miles (185 km) southwest of Amman and about 60 miles (97 km) northeast of Aqaba.

For Those Who Love Adventure

4. Mauvais must be running scared. He's been spotted trying to sell the lamp to a trader in Omdurman, Sudan. When Carolina gets there, she finds that Mauvais has taken the lamp to Meroë, capital of the ancient kingdom of Kush. It's a ruin now, full of steep-sided pyramids and mysterious stone buildings. Carolina takes a slow truck to a town about 25 miles (40 km) southwest of the ruin and about 95 miles (153 km) northeast of Omdurman.

5. Too late! Mauvais has fled across the desert by camel caravan. His destination is a town in west Africa. In the 1300s, it was a great center for trade and learning. Then war and drought brought about its decline. It's on the Niger River, about 435 miles (700 km) northeast of Bamako, Mali, and about 410 miles (660 km) northwest of Niamey, Niger. Carolina follows in a rented plane.

6. Mauvais has left the caravan and taken a plane to a city in France. From its location, Carolina guesses that he is going to the famous cave of Lascaux, about 35 miles (56 km) south. This underground "art gallery" was painted by Stone Age people about 25,000 years ago. Carolina tries to trail Mauvais in the city, about 110 miles (177 km) northeast of Bordeaux.

7. Carolina finds that Mauvais has bought a plane ticket for Manaus, Brazil. "He's got a hideaway in the Amazon rain forest," she thinks. She goes to Manaus, then up the river to a town about 325 miles (523 km) west and a little south. Sure enough, he's been there—but he's gone, further up the river and into Peru.

For Those Who Love Adventure

8. Carolina stops to catch her breath. But there's not a lot of breath to catch 8,000 feet above sea level. She's at an ancient Incan city in Peru, about 50 miles (80 km) northwest of Cusco, and about 300 miles (483 km) southeast of Lima. Legend has it that Incan leaders fled here after the Spanish conquest in the 1500s.

9. In Lima, Carolina learns that Mauvais has rented a boat. He was seen sailing southwest across the Pacific. "If I were trying to hide," she thinks, "I would go to that island with the great Polynesian stone heads. It's the most remote spot on earth—1,500 miles (2,415 km) from any other land." The island is about 2,300 miles (3,703 km) northwest of Santiago, Chile.

10. On the island, Carolina learns that Mauvais has taken the lamp to Alaska. She charters a boat, and then a plane. She catches him in Bettles, a little town south of Gates of the Arctic National Park. When she rubs the lamp, a genie appears. He whisks Carolina and her captive away to a city about 180 miles (290 km) southeast of there. Mauvais is in jail. The genie is back in the lamp—and the lamp is in a museum where it belongs.

Can you identify the cities and sites suggested by the clues? Write their names on the lines below.

1. _____   6. _____
2. _____   7. _____
3. _____   8. _____
4. _____   9. _____
5. _____   10. _____

Natural Wonders

"Moray" Murray, the ace counterfeiter, had a reward out for his capture—$100,000 in *real* money. I was determined to collect that reward and Moray knew it. He loved to give me the business about it, too. Matter of fact, that's how I finally caught him. He sent me a letter with a picture of himself: "Having a wonderful time, wish you were here." Ha-ha! But the blue glow surrounding him in the picture—I'd seen it before. I'm a nature photographer as well as a detective. Somewhere in my travels, I'd seen that blue glow. I began to go through my collection, photograph by photograph.

1. One set of photos was as red as Moray's was blue. They showed a huge rock in the middle of the Australian desert, 2,844 feet (867 m) tall. The Aborigines of Australia call it Uluru. It's in a national park of that name, about 210 miles (338 km) southwest of Alice Springs and about 800 miles (1,287 km) northwest of Adelaide.

2. Here were more red pictures—fiery red. They were of an erupting volcano in Hawaii. It's in Hawaii Volcanoes National Park, about 20 miles (32 km) from Mauna Loa. I photographed the geysers of lava while standing on the rim of the crater.

3. Speaking of geysers, I found my photos of the "original" that gave them their name. It means "gusher" in the Icelandic language. It shoots up its boiling water 195 feet (59 m) in the air about 50 miles (80 km) northeast of Reykjavik.

4. The next photos were mostly green and brown. They were pictures of the world's largest banyan tree. It's located in an island nation just southeast of India. Banyans keep putting down new trunks, so that a single tree can make a whole forest. This one has nearly 4,000 trunks.

5. *Here* was something blue. I had "shot" these beautiful limestone hills in southern China. They were near a city whose name starts with "G" and has six letters. It's in Guangzhi Province, about 240 miles (386 km) northwest of Guangzhou. But these were just the pale blue of sky and reflected sunlight. Moray's picture was *really* blue.

6. And here was some blue-green sea water—lots of it. I'd taken these pictures in Canada, where I'd gone to photograph the highest tides in the world. It's a bay between the provinces of New Brunswick and Nova Scotia. The tides in the narrowest part of the bay sometimes rise 50 feet (15 m).

7. Here was a photograph of more water, but the water was silvery rather than blue. It's a waterfall in Caraima National Park in Venezuela, about 410 miles (660 km) southeast of Caracas. It's just a little trickle of a falls on the Churún River—but it falls 3,212 feet (979 m)!

8. I gazed for a long time at pictures of my favorite mountain. It's topped with snow all year round, even though it's practically on the equator. It's in a national park about 310 miles (500 km) east and a little south of Mwanza, Tanzania, and about 125 miles (201 km) southeast of Nairobi, Kenya.

9. Now, here was water as blue as it gets. It was a picture of the deepest lake in the United States. It's in the crater of an extinct volcano, Mount Mazama. It's in a national park, about 90 miles (145 km) southeast of Eugene, Oregon.

10. I was feeling blue myself by then. If I couldn't figure out where Moray's card came from, I might just as well crawl into a cave. Wait—that was it—a cave! The picture had been taken in the Blue Grotto. It's a cave on an island just about 20 miles (32 km) south of Naples, Italy. The reflection of the sun on the sea inside the cave turns the walls a deep blue color.

   I caught a flight to Naples and brought Moray Murray back in handcuffs.

Can you identify the 10 natural wonders or their locations as suggested by the clues? Write their names on the lines below.

1. _____   6. _____
2. _____   7. _____
3. _____   8. _____
4. _____   9. _____
5. _____   10. _____

Celebrity Landmarks

Some buildings, natural features, and works of art are so well known that many people around the world recognize them instantly—even if they have never seen them personally. See how well you can do at identifying the locations of these "celebrity landmarks."

1. This bell tower was built for a cathedral in Italy. It started to tilt even before it was finished, because the soil it was built on is too soft. It's in a city about 42 miles (68 km) west and a little south of Florence. The "Leaning Tower" has been a landmark for so long that the city has no wish to straighten it out.

2. This statue of a mermaid looks out to sea in the harbor of a European capital. It's "The Little Mermaid," a character in a tale by Hans Christian Andersen. He was born in Odense, this country's second-largest city, about 90 miles (145 km) southwest of here.

3. Here is a statue of Jesus—one of perhaps millions in the Christian world. But the "Christ of the Andes" stands 12,674 feet (3,863 m) above sea level, on the border betweezn Chile and Argentina. It represents a declaration of peace between the two countries. It stands in Uspallata Pass, about 90 miles (145 km) northeast of Santiago, and about 35 miles (56 km) northwest of an Argentine city whose name begins with "M" and has seven letters.

4. Speaking of religious art—this sculpture is called Daibutsu. That's Japanese for "Big Buddha." It's the largest free-standing bronze statue in the world. It's in a Tokyo suburb whose name has eight letters, beginning with "K." You'll find it on the Sagami Sea, about 30 miles (48 km) southwest of downtown Tokyo.

World Geography Mysteries
© The Learning Works, Inc.

Celebrity Landmarks

5. What is one of the world's most beautiful buildings? Many people would name the Dome of the Rock. It's a Muslim shrine in this city that is holy to three religions. It is built over the spot where, Muslims believe, the prophet Muhammad rose to heaven. The city is in Israel, about 75 miles (121 km) southeast of Haifa, and about twice that distance north and a little east of Elat.

6. Or, perhaps the world's most beautiful building is the Taj Mahal. This great white tomb was built by an emperor of India for his empress. It's in a city about 105 miles (169 km) southeast of New Delhi, and about 135 miles (217 km) east and a little north of Jaipur.

7. These little vehicles were once common in American cities. Today, they are found in only one. They're called "cable cars" because they're powered by a cable that runs under the street. The city is in California, on a bay with the same name as the city, about 50 miles (80 km) northwest of San Jose.

8. The Parthenon stands on a hill called the Acropolis—as it has for nearly 2,500 years. The people of this city built it to honor the goddess for whom the city was named. It's in Greece, a little more than 40 miles (64 km) east of Corinth, and about 190 miles (306 km) west and a little south of Izmir, Turkey.

Celebrity Landmarks

9. Australians are proud of this white building. It stands on the waterfront of its largest city and looks like the shell of some giant sea creature. In fact, it's an opera house. You will find the city on a bay of the Pacific Ocean, about 425 miles (684 km) northeast of Melbourne.

10. We started this tour with a tower; let's end it with another. Its city has a lot of famous landmarks, some of which are nearly a thousand years old. But this tower, built for a world's fair in 1889, is probably its most famous symbol. Here's a hint: It was designed by an engineer named Gustave Eiffel. It's about 210 miles (338 km) southeast of London, England, and about 540 miles (869 km) southwest of Berlin, Germany.

Can you identify the locations of these 10 landmarks? Write the names of the cities on the lines below.

1. _____   6. _____
2. _____   7. _____
3. _____   8. _____
4. _____   9. _____
5. _____   10. _____

World Geography Mysteries
© The Learning Works, Inc.

Wildlife Scavenger Hunt

Congratulations! You've been invited to take part in the first occasional Wildlife Scavenger Hunt. All you have to do is photograph the 10 listed animal species in the wild. Just to be helpful, we've provided clues to places around the world where you might find each of these animals. Most of the locations are national parks. All you need is your camera and about $30,000 for air fare and expenses. Ready to go?

1. **Bengal Tiger.** This is an endangered species. It once ranged across much of Asia, but fewer than 1,000 may now remain in the wild. About 100 of them are in a park in India. The name of the park begins with a "C." It's in Uttar Pradesh state, about 115 miles (185 km) northeast of New Delhi.

2. **Mountain Gorilla.** This animal also is endangered. Its habitat is the highlands of eastern Africa. One gorilla refuge is in a national park in Congo, named for the mountain range where it is located. You'll find it in the eastern part of the country, on its border with Uganda and a little north of Rwanda. Lake Edward is just south of the park.

3. **Bighorn Sheep.** This is not an endagered species. It's hard to find, but that's only because its habitat is remote. It likes high mountains where most predators can't reach it. One place to find it is in Canada's oldest national park. It's in the province of Alberta, about 65 miles (105 km) northwest of Calgary. You'll want to check out beautiful Lake Louise.

Wildlife Scavenger Hunt

4. **Giant Tortoise.** An island group is the home of this 500-pound reptile, first described by Charles Darwin in 1835. These islands are the home of many species found nowhere else. They're in the Pacific Ocean, about 600 miles (966 km) off the coast of Ecuador.

5. **Golden Tubastrea.** This beautiful animal is not a bird. It's not a mammal. It lives in the sea, but it's not a fish. Oh, did I tell you you'll need a boat and scuba-diving equipment on this scavenger hunt? Sorry, it must have slipped my mind. The golden tubastrea is a species of sea anemone. When it dies, its skeleton becomes part of a coral reef. You'll find it in the world's biggest coral reef. It extends for about 1,250 miles (2,013 km) off the northeast coast of Australia.

6. **African Elephant.** This species is endangered in some parts of Africa, but its numbers are increasing in others. One place where it thrives is in a large national park in northeastern South Africa, on the border with Mozambique. This park is a protected habitat for 140 species of mammals, as well as hundreds of birds, reptiles, and other types of animals.

7. **Pale-Throated, Three-Toed Sloth.** It's easy to observe this animal—when you can find one. That's because it moves so slowly. One place to find it is in a national park in Costa Rica. This is a small country, but its rain forest is home to 1 out of every 25 species of land animals in the world. You'll find the park less than 15 miles (24 km) north of San José, the country's capital.

8. **Common Vampire Bat.** Yes, this animal really does suck blood. But don't worry, it rarely attacks human beings. You'll find it not in Transylvania, but in the Americas. One likely place to spot it is in a national park in the Amazon rain forest of Brazil. It's named for a river (*rio,* in Portuguese) that flows through it. You'll find it about 165 miles (266 km) northwest of Manaus.

9. **Burchell's Zebra.** This will be your last African stop. You'll find this species and many others in what may be the greatest wildlife preserve anywhere in the world. It's a national park in Tanzania, on its border with Kenya, east of Lake Victoria and west of Lake Natron. This grassland is one of the last places in Africa where you can still observe large animal migrations.

10. **Komodo Dragon.** This isn't a real dragon, of course. (Have you ever seen one of those?) It's a species of lizard—the largest in the world. It's found only on an island in Indonesia and on a neighboring island. You'll find about 1,000 of them in a national park named for the island. It lies between the islands of Sumbawa and Flores. The city of Surabaya, on the island of Java, is about 475 miles (765 km) west and a little north.

Can you identify these national parks and wildlife preserves? Write their names on the lines below.

1. _____   6. _____
2. _____   7. _____
3. _____   8. _____
4. _____   9. _____
5. _____   10. _____

A World Full of People

The World of Music and Dance

When Lamont Warwick announced that he would be writing a book about music and dance around the world, I was instantly on alert. The musical world knows Lamont as its leading critic. I knew him as something more—or less. For Lamont was a spy. He would sell any country's secrets to the highest bidder. I signed on as his photographer. I was certain that somewhere on his trip around the globe, I would catch him in the act.

1. Lamont's first stop was at a jazz festival. Jazz began in the United States, but this great festival is held in Switzerland. It is in a city on Lake Geneva, just 13 miles (21 km) southeast of Lausanne and a little more than 40 miles (64 km) southwest of Bern. During the Cold War, this area was a hangout for spies. I heard some swinging tunes, but I saw no documents change hands.

2. From there, Lamont went to hear an opera in Mozart's hometown. It's in Austria, about 300 miles (483 km) northeast of our first stop. It's also about 155 miles (250 km) southwest of Vienna, for many years the "capital" of European classical music.

3. I accompanied Lamont to see the Bolshoi Ballet. *Bolshoi* means "big" in Russian. The word refers to the theater where the company dances, but it might well refer to their reputation. Are people in this country born dancing, or what? The city we were in is big, too. It's about 230 miles (370 km) northeast of Smolensk, and about 250 miles (402 km) southwest of Nizhniy Novgorod.

The World of Music and Dance

4. Next, we caught a plane for India. We took in a performance of *bharata natyam*—a form of dance that is thousands of years old. It was originally a religious dance, performed in temples. The dances enact legends of the Hindu gods and goddesses. We saw it in a city in southern India, about 520 miles (837 km) southeast of Bombay, and about 960 miles (1,546 km) southwest of Calcutta.

5. Our next stop was on an island in Indonesia, just east of Java. The people of this island have an ancient tradition of folk dance. The dances act out stories based on legends or historical events. Every movement helps to tell the story, even the wiggle of a finger.

6. I was beginning to think I had made a mistake. I hadn't seen Warwick do anything suspicious. But then he made a sudden change of plans. We headed for Africa. In Ghana, he met and talked with an Ashanti master drummer. We attended one of his classes. The complex rhythms he produced on his drums were almost like melodies. We were in a city about 125 miles (201 km) northwest of Accra, and about 190 miles (306 km) southwest of Tamale.

7. Something indeed was up. Warwick had never had any interest in "world music"— so why was he taking me about 350 miles east and a little south into Nigeria? We were in one of Africa's largest cities to hear a performance by King Sunny Ade. The city was on the sea, about 320 miles (515 km) southwest of the new capital, Abuja.

The World of Music and Dance

8. Things were getting stranger. We went next to England. But we were not attending a concert in London. Warwick wanted to visit the home town of the Beatles, one of the most popular rock-and-roll bands in history. He's always *hated* rock and roll! The city is on the Irish Sea, about 175 miles (282 km) northwest of London, and about 30 miles (48 km) west and a little south of Manchester.

9. Our next stop was in the West Indies. These islands are home to many different kinds of music and dance. Warwick danced the night away in a salsa club in the islands' largest city. It's about 500 miles (805 km) northwest of Kingston, Jamaica, and about 1,100 miles (1,771 km) northwest of San Juan, Puerto Rico.

10. The mystery was cleared up when we came home. Warwick met a woman at the Jacob's Pillow dance festival in Becket, Massachusetts. In a college town about 23 miles (37 km) east of there, near the Connecticut River, he handed her a suitcase. It was full of items he had purchased for her on the journey. "I'll go out of my way any time for you, Mom," he said fondly. It just goes to prove than not every critic is a scoundrel.

Can you name the nine cities and one island suggested by the clues? Write their names on the lines below.

1. _____ 6. _____
2. _____ 7. _____
3. _____ 8. _____
4. _____ 9. _____
5. _____ 10. _____

The World of Fun and Games

Hi there, sports fans! I'm Jock Rooter, your guide to the world of sports. And I do mean world! I'm on my way to see an exciting team sport played. It's a game in which two animals are involved. Well, that's not quite true. One animal is involved. The other is *committed*. Do you know what sport I'm talking about, and where it is played? Come along and find out. On the way, let's visit some other sports locations around the world.

1. We'll start with the world's most popular sport—soccer. It's played in almost every country. Many nations played traditional types of kick-ball games. But soccer, as we know it, began in England. We're in South America, though. We're in the country that won the first World Cup, back in 1930. We're in the city where they won it, too. It's on the Rio de la Plata, about 125 miles (201 km) east and a little south of Buenos Aires, Argentina.

2. Personally, I love baseball. I'm visiting the town that has produced more major-league players than any city of its size in the world. It's in one of two countries on the West Indian island of Hispaniola, about 35 miles (56 km) east of Santo Domingo. Hey, baseball is a team sport played with two animals, isn't it? The ball is covered with horsehide—and you hit it with a bat!

3. Anyone for ice hockey? The game's "capital" is a French-speaking city about 325 miles (523 km) north of New York City. The first formal rules for ice hockey were drawn up here in the 1870s. The city is on the St. Lawrence River, about 100 miles (161 km) east of Ottawa.

The World of Fun and Games

4. I've always wanted to play golf in the country where it was invented. We're at the Royal and Ancient Golf Club of St. Andrews. The course was laid out in the 1700s. It's in Great Britain, but not in England. We're in a capital city on the Firth of Forth, about 45 miles (72 km) east and a little north of Glasgow.

5. Gymnastics isn't played with a ball or a puck. But what strength and grace it requires! We're in the capital of an Eastern European country. It's a small country, but its women's gymnastics teams are always among the best in the world. The city is about 130 miles (208 km) west of the Black Sea coast, and about 400 miles (644 km) southeast of Budapest, Hungary.

6. Here I am skiing in the Alps. I picked the perfect spot for it, a city that has twice hosted the winter Olympic Games. It's about 60 miles (97 km) south and a little west of Munich, though it's not in Germany. It's about 135 miles (217 km) east and a little south of Zurich, though it's not in Switzerland.

7. Speaking of winter sports—the canals of this European capital hardly ever freeze deeply enough for speed skating any more. Yet this is where the sport began. (It's gotten warmer since the 1800s.) We're about 220 miles (354 km) northeast of London, England, and a little more than 30 miles (48 km) northeast of The Hague.

The World of Fun and Games

8. Have you ever seen a rugby match? It's a game much like American football, only the players don't wear padding. It was invented in England, but it may be most popular in this city. We're near the southern tip of Africa, about 30 miles (48 km) northeast of the Cape of Good Hope.

9. Now let's go to Japan to see a championship karate match. I bet you think karate was invented in Japan. Well, yes and no. It was invented on an island about 400 miles (644 km) off Japan's southern coast. It's the largest of the Ryukyu Islands, about 500 miles (805 km) east of the coast of China.

10. Here we are at last! We're near a city about 185 miles (298 km) northwest of its country's capital, Kabul, and about 420 miles (676 km) east of Mashhad, Iran. The game is called *buzkashi*. It's played by two teams on horseback. The object, as in football, is to carry something over a goal line. But what they're carrying isn't a ball—it's a goat! It's dead, of course. The winning team traditionally eats it for dinner. The game is dying out, too, but it is still sometimes played in this part of Asia.

Can you identify the cities and countries, and one island, that Jock visited? Write their names on the lines below.

1. _____   6. _____
2. _____   7. _____
3. _____   8. _____
4. _____   9. _____
5. _____   10. _____

Sid and Max, Together Again

"All right, I'm going to give you another chance," said film producer Sid Sinnema. "We're going to make a new movie of Lucy Maud Montgomery's classic book, *Anne of Green Gables*. Your job is to find the right location to shoot the film."

"You've asked the right guy, chief!" said Max Gofer. "Just let me get my passport and I'll be on my way."

1. "Hello, Sid!" Max called in on his cellular phone. "I think I've found just the place we want. It's gorgeous. I'm on a beautiful river. There are mountains fading away in the distance. The town is full of pagodas. It's about an hour's flying time west of Shanghai—"

    "Max," Sid interrupted. "You're in the wrong part of the world. We're not filming Laurence Yep's *Mountain Light*, or anything set in _____ ."

2. "Hello, Sid!" Max called in a week later. "Wail till you see this place! I'm in a city just about 100 miles (161 km) west of the highest mountain in the world. The establishing shots will be—"

    "—pretty, but a lot less spectacular than that," Sid said. "Max, forget mountains. Forget _____ . We're not filming Robert Roper's *In the Caverns of Blue Ice*."

3. "Sid, Max here," came a call a few days later. "I know what you said about mountains, but wait till I tell you! I'm in this pretty little country surrounded by Germany, France, Italy, and Austria. They speak a lot of languages here! It seems to be a wealthy country, so I think it will be a comfortable place for a film crew—"

    "—if we were filming yet another version of Johanna Spyri's *Heidi*," Sid finished the sentence. "Max, get out of _____ . I think the high elevation has made you light in the head."

4. "Sid, I've got just the place," Max called in. "It's a country on a peninsula. It's south of Albania, Macedonia, and Bulgaria. There are mountains here, but I'm in a town on the Aegean Sea."

   "You must think the project is *Morning of the Gods*, by Edward Fenton," Sid groaned. "Not _____ , Max. Think of someplace that's colder."

5. "Okay, you want colder?" said Max a few days later. "I'm freezing here. I'm near the North Sea coast, about 275 miles (443 km) due east of Birmingham, England. It's windy. There are lots of windmills here."

   "Right, Max," Sid said wearily. "You're in _____ . *Hans Brinker*, by Mary Mapes Dodge, is a great classic, but it's not the movie we're making!"

6. "Okay, I'm about 400 miles (644 km) northeast of the last place I called from," Max said the next day. "I'm in a harbor looking eastward at Sweden across a narrow strait. I'm—"

   "—in dire straits yourself," Sid said threateningly. "We're not shooting Lois Lowry's *Number the Stars*. Get out of _____ !"

7. "Boss, I'm in a country just southwest of Belgium," Max reported the next day. "I'm in a sidewalk cafe beside the river Seine. I can see the Cathedral of Notre Dame from here. Funny, no one here seems to know anything about football."

   "That's because you're in _____ ," said Sid. "Are you scouting locations for Ludwig Bemelmans' *Madeleine* stories, or what?"

Sid and Max, Together Again

8. "All right, I'm northwest of the last place," Max said. "This would be a great place for a movie, except the cars are being driven on the wrong side of the street. The people here even speak English."

   "That's because you're in _____ ," Sid roared. "What do you think we're doing—another version of James M. Barrie's *Peter Pan*? Forget about it, Max. Come home. Take a vacation."

9. "Sid, I took your advice," Max phoned in. "I'm on a beach—can you hear the sea? I'm *near* home—only about 800 miles (1,287 km) southeast of Hollywood. But hey, this would be a great place for the movie—"

   "—if we were adapting it from *The Black Pearl*, by Scott O'Dell," Sid sighed. "But you're getting warmer. At least you're on the right continent."

10. "Okay, I'm in a place called Prince Edward Island," Max called in a few days later. "This could be it. There's a house here called Green Gables farmhouse. There's hundreds of visitors here. They've all read that book of yours. A lot of them think it was a true story and that it happened right here."

    "Wait right there until I get there," said Sid. "I wouldn't want you getting lost."

Can you name the 10 countries Max visited on his search for the movie location? Write their names on the lines below.

1. _____   6. _____

2. _____   7. _____

3. _____   8. _____

4. _____   9. _____

5. _____   10. _____

Let's Party!

"She likes festivals," Ray said. "She" was Stella Dorris, known on six continents for software piracy. Ray fed slides into the projector. "All over the world, she turns up at these big street parties. Next thing you know, the market is flooded with bootleg video games.

1. "Look at this," Ray said. "Chinese New Year, two Februaries ago. There she is, next to the guy in the lion suit. There are Chinese people all over the world who celebrate this holiday. Guess where the phony CDs turned up?"

    "I'd say in Canada," I said, studying the slide. "That's the city with the biggest Chinese community on the North American Pacific coast. It's about 75 miles (121 km) east of Port Alberni, and about 120 miles (193 km) northwest of Seattle, Washington."

2. "Right you are," said Ray. "Now look at this one." The slide showed colorfully dressed men riding on horseback through a dusty street. "This is a Muslim festival. It marks the end of the holy month of Ramadan. This was taken in northern Nigeria, where it's called—"

    "Sallah," I broke in. "I see her there in the crowd. That must be the city on the Hadejia River, about 220 miles (354 km) northeast of Abuja."

3. "Now, here she is at a *Carnaval* celebration." The slide showed great crowds of people dancing, and men playing steel drums. "This could be a lot of places in Central or South America—"

    "But it happens to be in Trinidad," I said. "I've been on that street. It's the capital, on the Gulf of Paria north of Venezuela."

4. "Right again," Ray said, clicking the projector. "Now, here she is in Mexico, last November 2."

   "Of course, the Day of the Dead." I waved a hand at the skeletons. "It's a way that people remember their dead and laugh in the face of death. I'd say that was taken in a city with a four-letter name, about 200 miles (322 km) northwest of Mexico City and about 110 miles (177 km) northeast of Guadalajara."

5. "Here she is in Japan," Ray said. The slide showed people carrying a brightly painted float through the street.

   "Of course, it's the Gion festival," I said. "It's held every July to honor the gods for ending a plague years ago. That's in Japan's old capital, about 230 miles (370 km) southwest of Tokyo, and about 28 miles (45 km) northeast of Osaka."

6. "And here she is in Germany." Ray spotlighted Stella in a crowd at a long wooden table laden with food and beer. "They call it Oktoberfest, but this was taken in September."

   "Right," I said. "It's an old harvest festival. I'll bet this was taken in that big city on the Isar River, about 120 miles (193 km) southeast of Stuttgart."

7. "This slide was taken in India," Ray said. "Look, they're carrying pictures of Mahatma Gandhi."

   "And I'll bet it was October 2nd—his birthday. That's in the capital. It's about 720 miles (1,160 km) northeast of Bombay."

8. "Okay, okay," Ray said. "Look, this is in Israel. It was taken just this February. Must be some Jewish religious holiday."

   "It's in the Bible, but it's not exactly religious," I said with a smile. "It's Purim. It celebrates the deliverance of the Jewish people by Queen Esther, but it's mostly an occasion for dressing in costume and having fun. That must be in the big city on the coast, about 30 miles (48 km) northwest of Jerusalem."

9. "That's Stella, jumping over a fire." Ray pointed to the screen. It's a New Year's festival in Azerbaijan—"

   "—But they celebrate it in the spring," I said. "It's called *Novus Bayrami*. That must be in the capital, on the Caspian Sea, about 280 miles (451 km) east of Yerevan, Armenia."

10. "Okay, okay, so you know your geography," Ray said. "Tell me this, smart guy. It's August. Stella's been reported back in Canada. Where should the Bureau look for her?"

    "Why, at the Klondike Gold Discovery Day festival," I said. "It's held at a town in the Yukon Territory, on the Yukon River, about 265 miles (426 km) northwest of Whitehorse. Good luck!"

Can you name the cities where Stella attended the festivals? Write their names on the lines below.

1. _____   6. _____
2. _____   7. _____
3. _____   8. _____
4. _____   9. _____
5. _____   10. _____

Answer Key

Spying Out the U.S.A.
Pages 8–10
1. Indianapolis, IN
2. Texarkana, TX-AR
3. Tacoma, WA
4. St. Petersburg, FL
5. Boston, MA
6. Ukiah, CA
7. Rock Island, IL
8. Gallup, NM
9. Huntington, WV
10. Pittsburgh, PA

Canada, East to West
Pages 11–13
1. Halifax
2. Québec
3. Montréal
4. Ottawa
5. Toronto
6. Winnipeg
7. Regina
8. Calgary
9. Vancouver
10. Whitehorse

Deeper Into Mexico
Pages 14–16
1. Tijuana
2. Chihuahua
3. Guadalajara
4. Acapulco
5. Mexico City
6. Puebla or Puebla de Zaragosa
7. Veracruz
8. Cancún
9. Oaxaca
10. Juárez or Ciudad Juárez

Mystery Cruise
Pages 17–19
1. Nassau, Bahamas
2. Havana, Cuba
3. Guatemala City, Guatemala
4. San Salvador, El Salvador
5. Managua, Nicaragua
6. Panama City, Panama
7. Kingston, Jamaica
8. Port-au-Prince, Haiti
9. Santo Domingo, Dominican Republic
10. San Juan, Puerto Rico

Down South America Way
Pages 20–22
1. Cusco or Cuzco, Peru
2. La Paz, Bolivia
3. Guayaquil, Ecuador
4. Maracaibo, Venezuela
5. Cayenne, French Guiana
6. Rio de Janeiro, Brazil
7. Brasília, Brazil
8. Buenos Aires, Argentina
9. Santiago, Chile
10. Punta Arenas, Chile

Answer Key

The Countess's Grand Tour
Pages 24–26
1. Reykjavik, Iceland
2. Dublin, Ireland
3. Paris, France
4. Valencia, Spain
5. Geneva, Switzerland
6. Milan, Italy
7. Munich, Germany
8. Vilnius, Lithuania
9. Odessa, Ukraine
10. Istanbul, Turkey

African Safari
Pages 30–32
1. Addis Ababa, Ethiopia
2. Nairobi, Kenya
3. Zanzibar, Tanzania
4. Antananarivo, Madagascar
5. Johannesburg, South Africa
6. Kinshasa, Congo
7. N'Djamena, Chad
8. Accra, Ghana
9. Gao, Mali
10. Monrovia, Liberia

Half the World is Asia
Pages 33–35
1. Beijing or Peking, China
2. Guangzhou or Canton, China
3. Chongqing or Chungking, China
4. Ulaanbaatar or Ulan Bator, Mongolia
5. Irkutsk, Russia
6. Vladivostok, Russia
7. Seoul, South Korea
8. Osaka, Japan
9. Sapporo, Japan
10. Toshkent or Tashkent, Uzbekistan

Desert Lands
Pages 27–29
1. Cairo, Egypt
2. Jiddah, Saudi Arabia
3. Djibouti, Djibouti
4. Khartoum, Sudan
5. Algiers, Algeria
6. Baghdad, Iraq
7. Damascus, Syria
8. Jerusalem, Israel
9. Esfahan or Isfahan, Iran
10. Kabul, Afghanistan

Answer Key

Half of Asia is Southern Asia
Pages 36–38
1. Srinagar, India
2. Bombay, India
3. Dhaka, Bangladesh
4. Kathmandu, Nepal
5. Mandalay, Myanmar or Burma
6. Bangkok, Thailand
7. Ho Chi Minh City, Vietnam
8. Singapore, Singapore
9. Manila, Philippines
10. Jakarta, Indonesia

Mountains and Deserts
Pages 43–45
1. Gobi Desert
2. Ural Mountains
3. Atacama Desert
4. Atlas Mountains
5. Kalahari Desert
6. Hindu Kush Mountains
7. Sonora Desert
8. Harz Mountains
9. Ar Rub' al Khali
10. Mid-Atlantic Ridge

Lakes and Inland Seas
Pages 46–47
1. Caspian Sea
2. Great Bear Lake
3. Inland Sea
4. Baltic Sea
5. Lake Titicaca
6. Lake Tanganyika
7. Lake St. Clair
8. Lake Eyre
9. Aral Sea
10. Dead Sea

Lands Down Under
Pages 39–41
1. Sydney, Australia
2. Melbourne, Australia
3. Perth, Australia
4. Alice Springs, Australia
5. Port Moresby, Papua New Guinea
6. Suva, Fiji
7. Auckland, New Zealand
8. Nuku'alofa, Tonga
9. Agana, Guam
10. Mount Erebus, Antarctica

Answer Key

Around the World, Island to Island
Pages 48–50
1. Manhattan Island
2. Bermuda or Bermuda Islands
3. Malta
4. Sri Lanka
5. Sumatra
6. Tahiti
7. Pitcairn Island
8. Tierra del Fuego
9. Falklands or Falkland Islands
10. St. Helena

New on the Map
Pages 51–53
1. Ukraine
2. Armenia
3. Croatia
4. Slovakia
5. Belize
6. St. Kitts and Nevis
7. Suriname
8. Namibia
9. Eritrea
10. Kyrgyzstan

Follow the River
Pages 54–56
1. Euphrates
2. Niger
3. Mackenzie
4. Danube
5. Darling
6. Amazon
7. Zambezi
8. Ganges
9. Volga
10. Mississippi

Dead Reckoning
Pages 57–58
1. Florence, Italy
2. Mount Popocatépetl, Mexico
3. Pago Pago, American Samoa
4. Panama Canal, Panama
5. Asunción, Paraguay
6. Carthage, Tunisia
7. Shanghai, China
8. Liechtenstein
9. Serengeti Plain, Tanzania
10. Mount Fuji, Japan

A Capital Adventure
Pages 60–62
1. New York City, United States
2. London, England or United Kingdom
3. Rome, Italy
4. St. Petersburg, Russia
5. Mecca, Saudi Arabia
6. Bombay, India
7. Tokyo, Japan
8. São Paulo, Brazil
9. Mexico City, Mexico
10. Los Angeles, United States

Answer Key

Don't Forget Your Camera!
Pages 63–65
1. Teotihuacán
2. Cork
3. Salisbury
4. Venice
5. Luxor
6. Zimbabwe or Great Zimbabwe
7. Siem Reap
8. Zhangjiakou (or Kalgan)
9. Fuji or Mount Fuji
10. Orlando

For Those Who Love Adventure
Pages 66–68
1. Gilgit
2. Kota Kinabalu
3. Petra
4. Shandi or Shendi
5. Tombouctou or Timbuktu
6. Limoges
7. Tefé
8. Machu Picchu
9. Easter Island
10. Fairbanks

Natural Wonders
Pages 69–71
1. Ayers Rock
2. Kilauea
3. Geysir
4. Sri Lanka
5. Guilin
6. Bay of Fundy
7. Angel Falls
8. Mount Kilimanjaro
9. Crater Lake
10. Capri

Celebrity Landmarks
Pages 72–74
1. Pisa
2. Copenhagen
3. Mendoza
4. Kamakura
5. Jerusalem
6. Agra
7. San Francisco
8. Athens
9. Sydney
10. Paris

Answer Key

Wildlife Scavenger Hunt
Pages 75–77
1. Corbett National Park
2. Virunga National Park
3. Banff National Park
4. Galapagos Islands
5. Great Barrier Reef
6. Kruger National Park
7. Braulio Carrillo National Park
8. Jaú National Park
9. Serengeti National Park
10. Komodo Island National Park

The World of Music and Dance
Pages 79–81
1. Montreux
2. Salzburg
3. Moscow
4. Bangalore
5. Bali
6. Kumasi
7. Lagos
8. Liverpool
9. Havana
10. Northampton

The World of Fun and Games
Pages 82–84
1. Montevidéo, Uruguay
2. San Pedro de Macorís, Dominican Republic
3. Montréal, Canada
4. Edinburgh, Scotland
5. Bucharest, Romania
6. Innsbruck, Austria
7. Amsterdam, The Netherlands
8. Cape Town, South Africa
9. Okinawa
10. Mazar-e Sharif, Afghanistan

Sid and Max, Together Again
Pages 85–87
1. China
2. Nepal
3. Switzerland
4. Greece
5. The Netherlands
6. Denmark
7. France
8. England
9. Mexico
10. Canada

Let's Party!
Pages 88–90
1. Vancouver
2. Kano
3. Port of Spain
4. Léon
5. Kyoto
6. Munich
7. New Delhi
8. Tel Aviv
9. Baku
10. Dawson